Innovative Machine Learning Applications in the Aerospace Industry

Venkata Tulasiramu Ponnada
Collins Aerospace, USA

IGI Global
Scientific Publishing
Publishing Tomorrow's Research Today

Vice President of Editorial	Melissa Wagner
Director of Acquisitions	Mikaela Felty
Director of Book Development	Jocelynn Hessler
Production Manager	Mike Brehm
Cover Design	Jose Rosado

Published in the United States of America by
IGI Global Scientific Publishing
701 East Chocolate Avenue
Hershey, PA, 17033, USA
Tel: 717-533-8845
Fax: 717-533-7115
Website: https://www.igi-global.com E-mail: cust@igi-global.com

Library of Congress Cataloging-in-Publication Data

Names: Ponnada, Venkata, 1983- editor
Title: Innovative machine learning applications in the aerospace industry / edited by Venkata Ponnada.
Description: Hershey, PA : IGI Global Scientific Publishing, [2026] | Includes bibliographical references and index. | Summary: "The primary objective of the book is to provide a comprehensive and insightful exploration of how machine learning (ML) is transforming the aerospace sector. This book aims to showcase real-world applications and case studies that demonstrate the successful integration of machine learning in various aerospace processes, highlighting the tangible benefits and innovations achieved"-- Provided by publisher.
Identifiers: LCCN 2025001341 (print) | LCCN 2025001342 (ebook) | ISBN 9798369375259 hardcover | ISBN 9798369375266 paperback | ISBN 9798369375273 ebook
Subjects: LCSH: Aerospace engineering--Data processing--Technological innovations | Machine learning--Industrial applications
Classification: LCC TL563 .I55 2025 (print) | LCC TL563 (ebook) | DDC 006.3/1--dc23/eng/20250429
LC record available at https://lccn.loc.gov/2025001341
LC ebook record available at https://lccn.loc.gov/2025001342

British Cataloguing in Publication Data
A Cataloguing in Publication record for this book is available from the British Library.

Table of Contents

Preface x

Chapter 1
Innovative Applications of Machine Learning in Aerospace Design and Manufacturing 1
G. Boopathy, Vel Tech Rangarajan Dr. Sagunthala R&D Institute of Science and Technology, Chennai, India
Balaji Ganesan, Hindustan Institute of Technology and Science, India
P. Sivaprakasam, Addis Ababa Science and Technology University, Ethiopia
T. Kumaran, Acharya Institute of Technology, India

Chapter 2
Machine Learning-Based Solutions for Aerospace Engineering 43
G. Prasad, Chandigarh University, Punjab, India

Chapter 3
Transforming Aerospace Design and Manufacturing Through Machine Learning 57
Kishorebabu Dasari, Keshav Memorial Institute of Technology, India
Sujana Parry, MTAR Technologies Limited, India
Srinivas Mekala, KG Reddy College of Engineering and Technology, India

Chapter 4
Next-Generation Computer Vision for Autonomous Drone Navigation: Cutting-Edge Techniques and Real-World Applications 87
Chikesh Ranjan, National Institute of Technology, Rourkela, India
Jonnalagadda Srinivas, National Institute of Technology, Rourkela, India
P. S. Balaji, National Institute of Technology, Rourkela, India
Kaushik Kumar, Birla Institute of Technology, India

Chapter 5
Multi-Scene Recognition in Single Aerial Images Using CNN 111
G. Ananthi, Mepco Schlenk Engineering College, India
S. Mehala Shevani, Mepco Schlenk Engineering College, India
P. Priyadharshini Devi, Mepco Schlenk Engineering College, India

Chapter 6
Role of Machine Learning in Computational Fluid Dynamics 135
G. Prasad, Chandigarh University, Punjab, India
Snehal Malik, Chandigarh University, Punjab, India
Aadya Gupta, Chandigarh University, Punjab, India
Yash Nigam, Chandigarh University, Punjab, India

Chapter 7
Investigation of Computer Vision and Machine Learning to Enhance Quality Control Processes in Aerospace Manufacturing: Innovative Machine Learning Applications in the Aerospace Industry 161
Dhirendra Patel, Amity University, Greater Noida, India
M. L. Azad, Amity University, Greater Noida, India

Chapter 8
Retracted Coaching Communication: A Conceptual Framework for Coaching Clients Toward More Emotionally Intelligent Intrapersonal.............................. 197
Dinorah Nieves, Rocky Mountain College of Art and Design, USA

Chapter 9
Catalyzing Aerospace Progress Machine Learning Strategies for Emerging Economics 225
Sreelekha Paul, Brainware University, India
Dipanjan Das, Brainware University, India

Compilation of References 243

About the Contributors 277

Index 285

Detailed Table of Contents

Preface x

Chapter 1
Innovative Applications of Machine Learning in Aerospace Design and Manufacturing 1
G. Boopathy, Vel Tech Rangarajan Dr. Sagunthala R&D Institute of Science and Technology, Chennai, India
Balaji Ganesan, Hindustan Institute of Technology and Science, India
P. Sivaprakasam, Addis Ababa Science and Technology University, Ethiopia
T. Kumaran, Acharya Institute of Technology, India

This chapter explores how machine learning (ML) is revolutionizing aerospace design and manufacturing, highlighting how it may improve operational efficiency, safety, and engineering precision. It describes how ML technologies enable smarter design, manufacturing optimization, and superior quality assurance in aerospace applications by discussing both historical and modern developments in the field. ML greatly enhances aerodynamic design, improves structural analysis, and speeds up computational fluid dynamics (CFD) simulations by using predictive algorithms and analyzing large datasets. It also explores the legal framework governing machine learning in the aerospace industry by tackling issues including data management, integration difficulties, and ethical concerns. This chapter provides a thorough review of current machine learning applications, new developments, and possible advancements in aerospace technology.

Chapter 2
Machine Learning-Based Solutions for Aerospace Engineering 43
G. Prasad, Chandigarh University, Punjab, India

The incorporation of machine learning (ML) in aircraft engineering has transformed the design, analysis, and operation of intricate aerospace systems. This study examines the present and developing applications of machine learning techniques in critical domains like aircraft design optimisation, defect detection and diagnostics, flight control systems, and predictive maintenance. Utilising extensive information from simulations, sensors, and real-time operations, machine learning models facilitate more efficient decision-making, improved system reliability, and decreased operational costs. Moreover, progress in deep learning, reinforcement learning, and neural networks is being progressively utilised for applications spanning aerodynamic modelling to autonomous flight control. This study emphasises the difficulties related to data quality, interpretability, and model validation in safety-critical aircraft contexts.

Chapter 3
Transforming Aerospace Design and Manufacturing Through Machine Learning 57
Kishorebabu Dasari, Keshav Memorial Institute of Technology, India
Sujana Parry, MTAR Technologies Limited, India
Srinivas Mekala, KG Reddy College of Engineering and Technology, India

Machine learning (ML) is driving change in aerospace design and manufacturing in order to increase efficiency, safety, and innovation. Complex designs and requirements required in aerospace industry. ML can improve design quality, support design thinking by generating multiple options, and enhance predictive analytics to detect potential problems. ML makes manufacturing processes smarter by predicting demand and managing resources effectively, increasing operational efficiency and improving product quality. ML helps to prevent errors and increase safety and security through predictive maintenance, non-destructive testing, and anomaly detection. ML can handle challenges such as data collection, interpreting in a safety-critical business context. ML offers great opportunities, from AI-powered innovation and human-machine collaboration to the potential impact of quantum computing in solving complex aerospace problems. Continued advances in machine learning will help to shape the future of aerospace, making machines more efficient, safer, and more innovative.

Chapter 4
Next-Generation Computer Vision for Autonomous Drone Navigation: Cutting-Edge Techniques and Real-World Applications 87
Chikesh Ranjan, National Institute of Technology, Rourkela, India
Jonnalagadda Srinivas, National Institute of Technology, Rourkela, India
P. S. Balaji, National Institute of Technology, Rourkela, India
Kaushik Kumar, Birla Institute of Technology, India

This chapter explores the cutting-edge advancements in computer vision technologies that are revolutionizing autonomous drone navigation. As drones are increasingly deployed in diverse applications such as mapping, surveillance, search and rescue, and delivery services, the need for precise and reliable navigation systems is paramount. The chapter begins with an overview of the fundamental principles of computer vision and its role in enabling autonomous navigation. The integration of machine learning algorithms, such as convolutional neural networks (CNNs) and deep learning frameworks, is discussed, highlighting their impact on enhancing the accuracy and efficiency of drone navigation. Furthermore, the chapter examines practical applications of these techniques, showcasing real-world scenarios where computer vision has significantly improved the performance of autonomous drones. Through this comprehensive examination, the chapter aims to provide valuable insights for researchers, engineers, and practitioners working to advance the field of autonomous drone navigation.

Chapter 5
Multi-Scene Recognition in Single Aerial Images Using CNN 111
G. Ananthi, Mepco Schlenk Engineering College, India
S. Mehala Shevani, Mepco Schlenk Engineering College, India
P. Priyadharshini Devi, Mepco Schlenk Engineering College, India

Scene recognition from aerial images is a primary application in utilizing high resolution satellite images. In the recent past, many studies were carried out in this field to classify an image to one scene category. But the real scenarios are different, they often contain more than one scene. Therefore, we explore an approach to recognize multiple scenes from a single aerial image. To do this, three different datasets namely UCM dataset, AID dataset and MAI dataset consisting of 2100, 10000, 3923 images respectively were used. Out of the three datasets, UCM and AID dataset are Single scene datasets and the MAI dataset is a multi-scene dataset. These datasets are utilized in two configurations namely UCM2MAI and AID2MAI. Prototype based memory network using different CNN baseline models is used as the backbone. First, the single scene datasets were used to extensively train different baseline models and store them as prototypes in the external memory. Experiments were conducted to analyse the performance of different baseline models.

Chapter 6
Role of Machine Learning in Computational Fluid Dynamics 135
G. Prasad, Chandigarh University, Punjab, India
Snehal Malik, Chandigarh University, Punjab, India
Aadya Gupta, Chandigarh University, Punjab, India
Yash Nigam, Chandigarh University, Punjab, India

The integration of machine learning (ML) with computational fluid dynamics (CFD) marks a significant advancement in the simulation and analysis of fluid flows. This chapter explores the synergistic role of machine learning in enhancing CFD methodologies, focusing on applications in modeling, optimization, and real-time analysis. Machine learning algorithms, particularly deep learning, offer powerful tools for identifying patterns and correlations within large datasets generated by CFD simulations. These algorithms can be trained to predict fluid behavior, accelerate simulation processes, and improve the accuracy of models by learning from empirical data. In modeling, ML techniques reduce the reliance on traditional empirical models, offering more precise and computationally efficient alternatives. Furthermore, ML-driven optimization techniques enhance the design process of fluid systems by enabling rapid evaluation of multiple design variables. Real-time data processing and analysis facilitated by ML also support adaptive control and decision-making in dynamic fluid environments.

Chapter 7
Investigation of Computer Vision and Machine Learning to Enhance Quality Control Processes in Aerospace Manufacturing: Innovative Machine Learning Applications in the Aerospace Industry 161
Dhirendra Patel, Amity University, Greater Noida, India
M. L. Azad, Amity University, Greater Noida, India

The integration of computer vision and machine learning (ML) in aerospace manufacturing has revolutionized quality control processes by enabling automated defect detection, predictive maintenance, and real-time monitoring. Traditional quality inspection methods often involve manual assessments, Computer vision systems, powered by deep learning algorithms, enhance defect detection accuracy by analysing images and sensor data, Machine learning further optimizes quality control by predicting potential failures through data-driven insights, reducing production downtime, and improving overall efficiency. This paper explores recent advancements in computer vision and ML for aerospace manufacturing, highlighting their impact on defect classification, anomaly detection, and process optimization. The study also discusses challenges such as data availability, model interpretability, and computational resource requirements. By leveraging intelligent automation, aerospace manufacturers can achieve higher reliability, cost efficiency, and compliance with stringent industry standards.

Chapter 8
Retracted Coaching Communication: A Conceptual Framework for Coaching Clients Toward More Emotionally Intelligent Intrapersonal.............................. 197
Dinorah Nieves, Rocky Mountain College of Art and Design, USA

Coaching has been described as a process of partnering with clients in a thought-provoking/creative way to inspire them toward maximum personal/professional performance. Often, that work comes down to two major tasks: helping the client communicate more effectively with key stakeholders in their lives and helping clients communicate more healthily with themselves. Yet, despite the salience of this statement, few academic pursuits work to ground the coaching process in the art of emotionally intelligent communications of the interpersonal and intrapersonal sort. This chapter proposes a research-based paradigm for theoretically and practically conceptualizing the coaching process in this way. Utilizing a pracademic approach, analyzing case study, and employing the use of valid and reliable scales, this author puts forth an argument for coaching as an effective process in facilitating the improved emotionally intelligent interpersonal and intrapersonal communication of coaching clients.

Chapter 9
Catalyzing Aerospace Progress Machine Learning Strategies for Emerging Economics 225
Sreelekha Paul, Brainware University, India
Dipanjan Das, Brainware University, India

The advancement in machine learning technology is impacting the aircraft industry globally as it develops at a very high rate. That is why more and more emerging states come to realize that aerospace innovation is the key to economic diversification and technical self-sufficiency and the path to national development. Machine learning has played an important role in enhancing design, safety, and operation efficiency; classified learning, learning from experience learning, neural networks, and local learning are quite suitable here. Some of the methods of avoiding such negative impacts include localization of data generating projects as well as cloud computing options that have been developed to suit the socio-economic conditions of emerging countries and; ethical AI governance frameworks.

Compilation of References 243

About the Contributors 277

Index 285

Preface

INTRODUCTION

The aerospace industry stands as one of humanity's most ambitious and innovative pursuits—where precision, performance, and safety are non-negotiable. For decades, this domain has pushed the boundaries of engineering and scientific exploration. Today, we find ourselves at the cusp of another transformative era, one defined by the integration of intelligent technologies—foremost among them, machine learning (ML).

Innovative Machine Learning Applications in the Aerospace Industry brings together a curated collection of research, case studies, and expert insights to showcase how machine learning is reshaping the aerospace landscape. From conceptual design and simulation to real-time maintenance and autonomous flight operations, ML is driving efficiencies, reducing costs, and enabling capabilities once thought impossible.

This book was conceived with several key goals in mind. First and foremost, it aims to educate by offering readers a solid grounding in the fundamental principles of machine learning, tailored specifically to the aerospace context. Secondly, it seeks to inspire—highlighting real-world examples and success stories that illuminate the enormous potential of these technologies. It also intends to inform, by examining current trends, emerging research, and future trajectories in ML development. At the same time, it aims to guide practitioners by providing practical frameworks, tools, and methodologies for implementation. Finally, and crucially, it works to advocate for responsible and ethical ML deployment—emphasizing the significance of safety, regulatory compliance, and human oversight.

The chapters herein are designed for a diverse readership. Aerospace engineers and professionals will find applications that resonate with their daily challenges. Data scientists and ML practitioners will gain insight into a domain where their tools are making a dramatic impact. Researchers and academics will discover rich intersections between disciplines, while students and educators will benefit from comprehensive material that bridges theory with practice. Industry leaders and policymakers will gain clarity on how machine learning is not just a technological add-on, but a strategic imperative. Even passionate aerospace enthusiasts will uncover new dimensions of innovation at the crossroads of flight and data science.

As the editor of this volume, I am proud to present this work as both a reflection of current advancements and a roadmap for future exploration. May it serve as a catalyst for learning, collaboration, and innovation in a field where precision meets possibility.

ORGANIZATION OF THE BOOK

In *Chapter 1: Innovative Applications of Machine Learning in Aerospace Design and Manufacturing*, we delve into how ML is revolutionizing the aerospace production pipeline. The chapter illustrates the evolution from conventional methods to smart systems that improve structural analysis, enhance design simulations, and ensure superior quality assurance. It also considers legal and ethical issues surrounding ML integration, creating a well-rounded foundation for understanding its role in modern aerospace systems.

Chapter 2: Machine Learning Based Solutions for Aerospace Engineering expands this exploration by surveying the transformative applications of ML across the aerospace engineering spectrum. From predictive maintenance and aerodynamic modeling to autonomous flight control, this chapter showcases how machine learning optimizes system reliability, cost-efficiency, and real-time decision-making—while also addressing the technical challenges of data quality and model interpretability in safety-critical environments.

In *Chapter 3: Transforming Aerospace Design and Manufacturing Through Machine Learning* continues this theme with a forward-looking perspective on innovation. This chapter emphasizes the generative potential of ML in design thinking, its predictive power in identifying system flaws early, and its contribution to enhanced product safety and process efficiency. The potential of future technologies such as quantum computing is also discussed, highlighting the possibilities for long-term transformation.

Chapter 4: Next-Generation Computer Vision for Autonomous Drone Navigation investigates the cutting-edge integration of ML and computer vision in drone technology. It explores how convolutional neural networks and deep learning frameworks are redefining autonomous navigation, especially in high-stakes environments like disaster response, mapping, and logistics. Through real-world case studies, the chapter underscores the operational impact of advanced visual processing systems.

Chapter 5: Multi-Scene Recognition in Single Aerial Images Using CNN transitions the discussion toward advanced imaging and classification challenges in aerospace. The chapter proposes a novel approach to multi-scene recognition using prototype-based memory networks, trained on various aerial datasets. It addresses the complexity of aerial imagery that often contains multiple scene types and demonstrates how CNN-based models can effectively interpret this layered data for mission-critical aerospace applications.

In *Chapter 6: Role of Machine Learning in Computational Fluid Dynamics*, we explore how ML is accelerating and refining computational fluid dynamics simulations. By integrating data-driven modeling and optimization techniques, ML enables more accurate predictions of fluid behavior, real-time adaptability, and reduced computational load—paving the way for smarter fluid system design and dynamic response mechanisms in aerospace engineering.

Chapter 7: Investigation of Computer Vision and Machine Learning to Enhance Quality Control Processes in Aerospace Manufacturing introduces us to the tangible impacts of AI in real-time quality assurance. With detailed coverage of how computer vision and ML enable automated defect detection and predictive analytics, this chapter demonstrates how intelligent automation ensures higher reliability, process optimization, and compliance with strict aerospace standards.

Chapter 8: Coaching Communication – A Conceptual Framework for Coaching Clients Toward More Emotionally Intelligent Intrapersonal offers a reflective shift, emphasizing the human dimension of innovation. This chapter introduces a research-based model for coaching emotional intelligence, particularly in self-communication, which has growing relevance in an AI-integrated workplace. Through a pracademic lens and case studies, it bridges the technical focus of the book with an essential humanistic perspective.

Finally, *Chapter 9: Catalyzing Aerospace Progress – Machine Learning Strategies for Emerging Economies*, we begin by examining how emerging nations are leveraging machine learning to drive aerospace innovation and, by extension, national development. The chapter outlines how ML contributes to design, safety, and operational efficiency, while also highlighting the need for ethical AI governance and context-specific adaptations such as cloud computing and localized data infrastructures.

CONCLUSION

As this volume demonstrates, the integration of machine learning and computer vision technologies within the aerospace domain is not simply a trend—it is a transformative force reshaping how we design, manufacture, operate, and maintain complex aerospace systems. Each chapter in this book contributes a vital perspective to this evolving narrative, showcasing the vast potential and practical impact of intelligent systems across various dimensions of aerospace innovation.

From theoretical explorations of design optimization and predictive maintenance to practical applications in aerial image recognition and autonomous navigation, the authors have illuminated how data-driven intelligence accelerates efficiency, safety, and innovation in emerging and established economies alike. Furthermore, the inclusion of interdisciplinary insights—such as the exploration of emotionally intelligent coaching models—emphasizes that as much as aerospace progress is driven by technology, it must also be guided by human-centered thinking and ethical considerations.

Together, these contributions form a robust compendium of contemporary thought and pioneering research. As the editor, I am confident that this collection will inspire further inquiry, collaboration, and implementation in the global aerospace community. May it serve as a foundational resource for researchers, engineers, practitioners, and policymakers committed to catalyzing progress at the intersection of machine learning and aerospace advancement.

Chapter 1
Innovative Applications of Machine Learning in Aerospace Design and Manufacturing

G. Boopathy
https://orcid.org/0000-0003-2515-7277
Vel Tech Rangarajan Dr. Sagunthala R&D Institute of Science and Technology, Chennai, India

Balaji Ganesan
https://orcid.org/0000-0001-5799-9423
Hindustan Institute of Technology and Science, India

P. Sivaprakasam
https://orcid.org/0000-0001-8082-8649
Addis Ababa Science and Technology University, Ethiopia

T. Kumaran
https://orcid.org/0000-0002-7165-0954
Acharya Institute of Technology, India

DOI: 10.4018/979-8-3693-7525-9.ch001

ABSTRACT

This chapter explores how machine learning (ML) is revolutionizing aerospace design and manufacturing, highlighting how it may improve operational efficiency, safety, and engineering precision. It describes how ML technologies enable smarter design, manufacturing optimization, and superior quality assurance in aerospace applications by discussing both historical and modern developments in the field. ML greatly enhances aerodynamic design, improves structural analysis, and speeds up computational fluid dynamics (CFD) simulations by using predictive algorithms and analyzing large datasets. It also explores the legal framework governing machine learning in the aerospace industry by tackling issues including data management, integration difficulties, and ethical concerns. This chapter provides a thorough review of current machine learning applications, new developments, and possible advancements in aerospace technology.

INTRODUCTION

Within the broader discipline of artificial intelligence (AI), machine learning (ML) is a separate field that focuses on creating statistical models and algorithms that let computer systems carry out tasks without the need for explicit programming instructions. These activities include identifying trends, making decisions, forecasting results, and enhancing performance via data-driven analysis and experiential learning. In contrast to conventional programming techniques, which necessitate the establishment of particular procedural rules, ML algorithms learn from data and gradually improve their working methods.

ML is quickly becoming a key element of innovation in the aerospace sector, propelling improvements in a number of areas, including design, manufacture, operations, and maintenance. Machine learning's significance stems from its capacity to handle and analyse enormous datasets at previously unattainable speeds, revealing insights that were previously impossible to obtain using traditional methods. This feature is especially important in the aerospace industry, where high levels of accuracy, dependability, and operational efficiency are required for the essential nature of operations and the complexity of systems.

The integration of ML within the aerospace sector presents numerous significant advantages:

Augmented Decision-Making: ML algorithms exhibit the capacity to analyse vast datasets in order to identify patterns and associations that may not be immediately perceptible to human engineers. This advancement promotes more prudent decision-making across various domains, including design optimization, predictive maintenance, and the formulation of operational strategies.

Enhanced Efficiency: ML lowers the time and resources needed for complicated processes like supply chain management, inspections, and simulations by automating them. Consequently, this leads to significant cost reductions and faster development cycles.

Elevated Safety and Reliability: The safety and dependability of aircraft systems are improved by the early detection of possible breakdowns and deviations made possible by machine learning's predictive capabilities. In a field where even, small mistakes can have disastrous results, this specific element is essential.

The application of ML in the aerospace sector is one aspect of a larger transformation taking place in the industry, where technological breakthroughs have continuously expanded the range of possible outcomes. This development began with the advent of digital computing technologies in the middle of the 20th century, which completely changed aircraft engineering and design by making it possible to perform complex computations and simulations that were previously thought to be impossible.

Advanced computational methods, such as computational fluid dynamics (CFD) and finite element analysis (FEA), were integrated into the aerospace industry during the 1980s and 1990s. By using these methods, engineers were able to create more accurate models and resemble the behaviour of aircraft systems, leading to more complex and efficient designs. However, these simulations frequently showed high processing requirements and long run times, which led to the creation of novel strategies to address the growing complexity of aircraft systems.

The aerospace sector has experienced considerable transformation since its establishment, influenced by technological innovations, market requirements, and shifts in regulatory frameworks. This transformation embodies a multifaceted interaction of historical achievements and present-day obstacles, which are essential for comprehending the current terrain of the industry. Technological breakthrough in hot air balloon technology for atmospheric manoeuvring. The Wright siblings created the first motorized airplane. The manuscript explores the evolutionary path of the aerospace ecosystem using network science techniques and economic ideas. The paper identifies recurring themes and similar characteristics in the development of aerospace manufacturing ecosystems in developed nations. (Jose Jr et al.,2020)

The work examines into the historical evolution of the aerospace industry, with a focus on quality management techniques, process integration approaches, and Industry 4.0 technology implementation to enhance productivity, sustainability, and competitive edge in aerospace operations (Pop et al.,2023). In the US, the aerospace industry moved from a monopolistic to a duopolistic structure, with the possibility of an oligopolistic market. This is a very significant, globally relevant industry driven by technological innovation and the complexities of market dynamics (Shah,2015). The development of the Brazilian aviation sector is examined using the MOIP and SIS theoretical frameworks. The importance of state involvement in achieving a position of worldwide competitiveness was also covered (Caliari et al., 2023). Analysis of the sectoral innovation system's evolutionary trajectory inside India's aviation industry. The Bangalore cluster, key stakeholders, technical developments, and changes in demand are highlighted (Sunil,2013).

A significant change occurred with the advent of ML in the early 21st century. ML was first used in specialized fields including anomaly detection, control systems, and signal processing. However, the variety of ML applications in the aerospace industry expanded quickly as computing power increased and large datasets became more accessible. These days, almost every aspect of aircraft engineering and manufacturing incorporates machine learning. Figure 1 shows the key turning points in the development of machine learning in the aerospace sector.

Figure 1. Key milestones in evolution of ML in aerospace

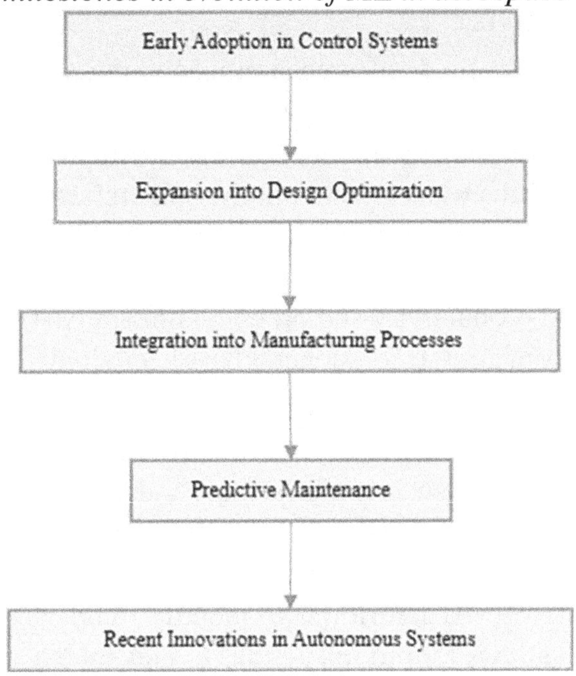

The aircraft industry was the first to use ML to improve flight control system optimization, especially when it came to handling dynamic and changing flying situations. These cutting-edge ML algorithms were designed to improve aircraft stability and control by integrating large sensor-derived information. These technologies' accomplishments laid the foundation for increasingly complex automation and autonomous aircraft applications. This period represented the initial exploration of machine learning's potential to enhance the functionality of critical aerospace systems.

ML began to play a crucial role in design optimization as computing power increased, particularly in reducing the computational load related to simulations like FEA and CFD. Engineers were able to explore a wider range of design options because ML models, which were trained on historical simulation information, demonstrated the capacity to predict results more quickly. Faster design cycles and more inventive aeronautical configurations were the outcomes of this development. For aeronautical engineering to become more effective and efficient, ML has to be integrated into design processes.

Particularly in the fields of precision machining and additive manufacturing (AM), ML has become an essential feature of aerospace manufacturing. To ensure the manufacture of high-quality outputs with fewer faults, ML algorithms were used to optimize manufacturing factors such as temperature and material flow rates. Additionally, by enabling real-time adjustments throughout the manufacturing process, these predictive models increased productivity and product uniformity. The performance and dependability of aircraft components have been significantly improved by the use of ML into manufacturing processes.

Predictive maintenance emerged as an essential use of machine learning, in which algorithms analyse aircraft sensor data to anticipate possible problems before they happen. While increasing safety and dependability, this approach drastically lowers unexpected downtime and maintenance costs. The ability of ML algorithms to identify trends and irregularities in large datasets allows for prompt actions to prevent expensive and dangerous issues. By ensuring that aircraft maintain ideal conditions for the duration of their working lives, predictive maintenance has become an essential part of modern aerospace operations.

Significant progress has been made in autonomous systems, such as autonomous airplanes and Unmanned Aerial Vehicles (UAVs), due to recent developments in machine learning. Even in dynamic and complex surroundings, these systems can navigate, avoid obstacles, and make decisions in real time by using ML algorithms. (Govindarajan et al., 2025) The development of completely autonomous flying, which has major implications for the commercial and military aircraft industries, depends on these skills. The continuous advancements in this field demonstrate how ML has the ability to revolutionize aerospace operations and push the envelope of what is possible.

It is expected that machine learning's importance in the aircraft industry will grow as it develops. In order to further enhance the capabilities of aircraft systems, future advances are probably going to focus on integrating ML with other emerging technologies, such edge computing and quantum computing.

THE ROLE OF MACHINE LEARNING IN AEROSPACE DESIGN

Aerospace design is undergoing a revolution due to machine learning, which is increasing productivity, accuracy, and creative potential in a wide range of applications. Its integration into design processes is radically transforming traditional methods, leading to significant advancements. By automating the design process, ML makes it easier to construct radar-absorbing structures for aerospace applications. This increases efficiency and maximizes the use of meta surfaces to manipulate electromagnetic waves (Anjana et al.,2024). The ML framework assesses the acceptable limits for digital designs related to composite materials used in aircraft. To evaluate and measure uncertainty, it combines failure analysis techniques with stochastic process simulations (Eskandariyun et al., 2024). ML tools like Physics-Informed Neural Networks and ODIL are developed for aerospace design, utilizing wind tunnel data on a speed bump model to bridge experimental and computational efforts (Humml et al., 2024).

By optimizing digital twins through continuous updates, combining sensor data, and using historical records, ML greatly advances the design of aircraft structures. This allows for accurate predictions, continuous monitoring, and well-informed decision-making throughout every step of the aircraft life cycle (Tavares et al., 2024). ML plays a crucial role in aerospace design by enhancing aerodynamics, CFD, aircraft design, aero acoustics, health state evaluation, component optimization, blade defect detection, and combustion in aerospace engineering (Wang et al., 2024). Shorten the time needed to develop airplanes by using an all-inclusive integrated design approach. Employ ML methodologies to ascertain the optimal architectural arrangements (Kalyan, 2024).

Through the automation of star classification and improvement of precision, ML greatly enhances design processes in the aerospace industry. This technical development changes the way that aircraft items are classified, which makes major advances in the field of aeronautical engineering possible (Vignesh et al., 2024). Aero-Engines AI utilizes ML models to evaluate airplane engines conceptually. Turbofan performance parameters are reliably predicted using ML algorithms (Tong, 2023). Aerodynamic inverse design is greatly accelerated by machine learning, which reduces computational costs by up to 80% through data-driven models that find connections between pressure distributions and geometric configurations. This increases the effectiveness of aerospace design (Shirvani et al., 2023). Design Engineers may now push the limits of creativity, effectiveness, and performance through ML, which is revolutionizing aircraft design. Figure 2 provides a detailed explanation of how it impacts each aspect of aeronautical design.

Figure 2. Influences of ML in various aspects of aerospace design

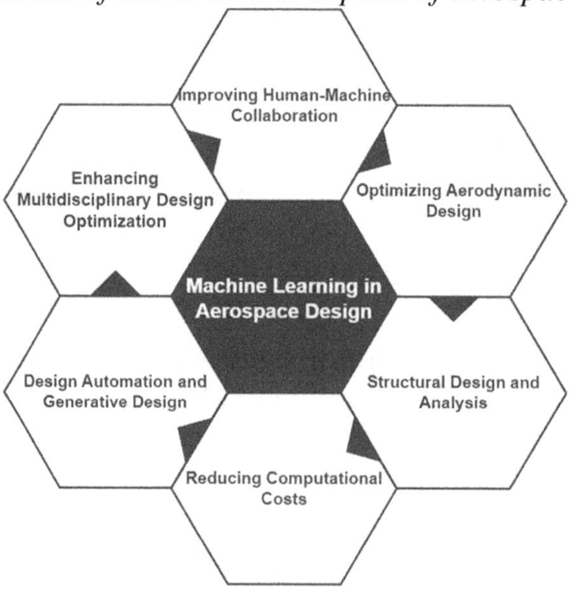

Optimizing Aerodynamic Design

Because it has a significant impact on an aircraft's efficiency, performance, and safety factors, aerodynamics is a crucial component of aeronautical engineering. Traditional aerodynamic design approaches mostly use CFD simulations, which are notably computationally intensive and time-consuming. The use of ML models, which are trained on large datasets that include previous simulations and experimental results, allows for precise aerodynamic performance prediction, significantly reducing the time needed for design iterations.

Through the implementation of ML algorithms, engineers are afforded the opportunity to investigate a broader spectrum of design variables, including wing geometries, airfoil configurations, and control surface arrangements, in order to ascertain optimal solutions that harmonize performance, fuel efficiency, and stability. Additionally, ML models make it possible to automate the design process, which enables the creation and evaluation of several design configurations in a substantially shorter amount of time than with conventional methods. More creative design solutions that are better suited to certain performance requirements are produced by this conceptual shift.

Aerodynamic design optimization is critical for performance improvement in a wide range of applications, including aircraft, propellers, and helicopters. CFD, ML and innovative design techniques are used in recent innovations to achieve significant improvements in effectiveness and efficiency. Aerodynamic form design heavily relies on CFD, particularly when working with intricate surfaces. When response surface modelling and a genetic algorithm are combined, the optimization process is greatly enhanced, making it possible to make flexible adjustments to design specifications and effectively handle complex geometries (Gao, 2023). The application of ML techniques, particularly in the field of turbulence modelling, has demonstrated significant promise for improving aerodynamic configurations. By using traditional models, researchers have achieved notable decreases in drag, demonstrating how the choice of turbulence model affects optimization outcomes (Bidar et al., 2024). The development of reduced-dimensional neural networks with multi-fidelity makes it possible to foresee optimal designs with great accuracy and at a significantly reduced computational cost. This design process is made more manageable by this methodology (Du et al., 2023).

Structural Design and Analysis

For aircraft engineering to achieve optimal performance and maximize fuel efficiency, lightweight and robust components must be developed concurrently. By predicting material properties, stress distribution, and failure modes under various loading scenarios, ML is increasingly being used to support structural design and analysis.

Engineers can improve the design of crucial parts like fuselages, wings, and landing gears by using ML models, which can analyse large datasets related to material behaviour and structural performance. These models play a vital part in identifying possible weak points and suggesting design changes that increase the structure's overall durability and strength without adding unnecessary weight. Furthermore, by predicting the performance and behavior of new composite materials, ML improves the material selection process and promotes innovation.

In the field of aerospace engineering, structural design and analysis are critical to ensuring the effectiveness, safety, and operational performance of aircraft components. Modern developments are focused on material selection, optimization techniques, and creative design frameworks that maximize structural integrity while minimizing weight. The choice of materials, including plastics and titanium alloys, has a big impact on how structurally sound aeronautical components are. For instance, titanium alloys are being researched for use in aviation wings because they offer better performance than traditional aluminum (Mishra et al., 2023).

Optimization techniques, such as lattice structures and topology optimization, are essential for improving the strength and durability of 3D printed parts and producing lighter, more effective designs (Raja et al., 2024). Cutting-edge computer methods, such as the Variational Asymptotic Method, enable the structural analysis of wing structures under dynamic loading circumstances with high efficiency and little mass loss while complying with safety requirements (Sarojini et al., 2022). The effectiveness of optimal configurations in real-world applications, like pressurized compartments, is supported by empirical validations, such as hydrostatic and airtight examinations (Zhizhong et al., 2023).

Reducing Computational Costs

Reducing the computing costs associated with simulations is one of the most prominent advantages of applying ML to aeronautical design. High-fidelity simulations, such those used in FEA and CFD, are essential for accurate design validation but also resource-intensive.

Without requiring extensive simulations, ML methods can serve as alternative models, providing quick approximations of simulation results. These alternative models can then be used to predict outcomes for new design configurations and are created using data from a limited number of high-fidelity simulations. This approach accelerates up the design process considerably, allowing engineers to explore a larger design area and iterate quickly. ML reduces the need for full-scale simulations, which saves time and cuts the cost of processing resources.

Aerospace engineering is undergoing a revolution because to ML, which significantly reduces the computing costs associated with traditional approaches. By utilizing ML techniques, researchers are developing models that streamline the design process, increase accuracy, and facilitate rapid iterations. Non-intrusive ML methods, like as auto encoders, are designed to create reduced-order models (ROMs) that enable quick aerodynamic analyses while requiring only a small portion of the resources compared to higher-level models (Moni et al., 2024).

Design Automation and Generative Design

In the field of design automation, where complex algorithms are used to automate labour-intensive and repetitive operations across the design lifecycle, ML plays a crucial role. One notable development in this area is generative design, where ML algorithms generate a wide range of design options based on predetermined goals and limitations.

In generative design, the engineer defines the desired performance parameters, including weight, strength, and aerodynamic efficiency, and the ML model then generates a range of possibilities for design that meet these predetermined criteria. By integrating input obtained from simulations or empirical testing, the algorithm iteratively improves the designs until it converges on the best option. Innovative design concepts that might not be readily apparent or logically observable to human designers might be investigated according to this methodology.

When combined with machine learning, generative design promotes the creation of innovative and efficient designs because it may identify solutions that balance conflicting goals, such minimizing weight while maintaining structural integrity. Additionally, it enables the production of highly customized parts that are tailored to certain operational needs.

The combination of generative design and design automation in aerospace engineering is revolutionizing the field by increasing productivity, lowering costs, and streamlining processes. This approach change makes it easier for engineers to focus on requirements articulation rather than traditional design techniques. Research highlights the need for a change in mindset in engineering by demonstrating how generative design refines structural configurations through the use of physics-informed simulations (Muelaner, 2024). Lockheed Martin's application of generative design to thin-walled buildings demonstrated improved configurations by breaking down complex designs into manageable parts, which improved performance and manufacturability (Action et al., 2024). Supercritical airfoil design is automated with the EvoGD methodology, which significantly reduces design time and increases aerodynamic efficiency by up to 7.1% when compared to traditional methods (Sun et al., 2023). It has been shown that using generative design to create three-dimensional models and apply intelligent parameter settings can increase the quality and efficiency of design for standard airplane structures (Chu et al., 2023).

Enhancing Multidisciplinary Design Optimization (MDO)

Aerodynamics, structural integrity, propulsion systems, and control mechanisms are just a few of the many fields that make up aerospace design, which is inherently interdisciplinary. The goal of MDO is to simultaneously improve each of these aspects rather than treating them separately.

It is feasible to optimize designs more thoroughly by combining and evaluating data from MDO. Using ML algorithms, which can find connections and trade-offs among many design elements, engineers can arrive at optimal solutions that consider the interdependencies present in multiple domains. Better-balanced solutions arise from machine learning's ability to bring attention to these choices early in the design process. For example, a modification meant to enhance aerodynamic performance might have the opposite effect on structural robustness.

Additionally, machine learning-powered MDO tools can quickly evaluate a large number of design options, giving engineers a comprehensive understanding of the design environment and assisting them in making smart choices. As a result, more robust and efficient aerospace designs that meet the demanding specifications of modern aviation are created.

Improving MDO in aircraft design requires combining several technical fields in order to increase productivity and performance indicators. The importance of comprehensive frameworks and approaches that facilitate this integrative process is highlighted by recent developments. By efficiently combining structural, thermal, and acoustic assessments, the Multidisciplinary Structural Analysis and Design Optimization (MSADO) framework maximizes spacecraft operational performance in a variety of environmental scenarios (Borwankar et al., 2024). Due to budgetary constraints related to launch operations, the use of high-fidelity CFD approaches for the optimization of aerodynamic designs is becoming in popularity (Adimurthy et al., 2023). It is critical to address uncertainty in design processes because they add complexity and restrict computational efficiency. In order to increase reliability in MDO, modern technologies are focused on quantifying these uncertainties (Hajela et al., 2023).

Improving Human-Machine Collaboration

Apart from enabling automation in specific aspects of the design process, ML greatly enhances human-machine interaction, allowing engineers to focus on critical decision-making and creative problem-solving. ML algorithms are able to analyse large datasets and provide information that helps designers make wise decisions.

To find trends and best practices, ML can be used, for example, to analyse performance metrics, failure reports, and historical design records. This information can be shared with designers as guidelines or design principles, which will help them avoid common mistakes and adopt proven practices. Additionally, machine learning-powered technologies can speed up the development of design concepts, allowing engineers to quickly assess the feasibility and effectiveness of different concepts.

To further improve human-machine collaboration in the field of aircraft design, ML requires a greater degree of interaction between human operators and AI systems. By focusing on interactive and explicable systems, user interfaces, and design techniques, this joint effort can be further streamlined. Productivity can be increased by deploying collaborative robots and ML agents, although traditional approaches usually lead to autonomous rather than cooperative behaviour. By enabling people to modify AI behaviours, iterative, mixed-initiative design concepts promote enhanced collaboration (Paleja et al., 2024). The PRODEC methodology optimizes performance in critical environments like aerospace by fusing artificial intelligence with human-centered design concepts. This approach promotes trust. It emphasizes how important human-in-the-loop simulations and scenario-based design are to improving AI systems (Boy & Morel, 2022). Human considerations must be incorporated into the design process in order to promote productive collaboration, and human-in-the-loop approaches make it easier for AI systems to be tailored to the unique requirements of users (Gómez-Carmona et al., 2024).

MACHINE LEARNING IN AEROSPACE MANUFACTURING

By introducing cutting-edge methods and procedures that improve productivity, quality, and creativity, ML is completely changing the aerospace manufacturing industry. Figure 3 provides a detailed analysis of the role of machine learning in this area.

Figure 3. Influences of ML in various aspects of aerospace manufacturing

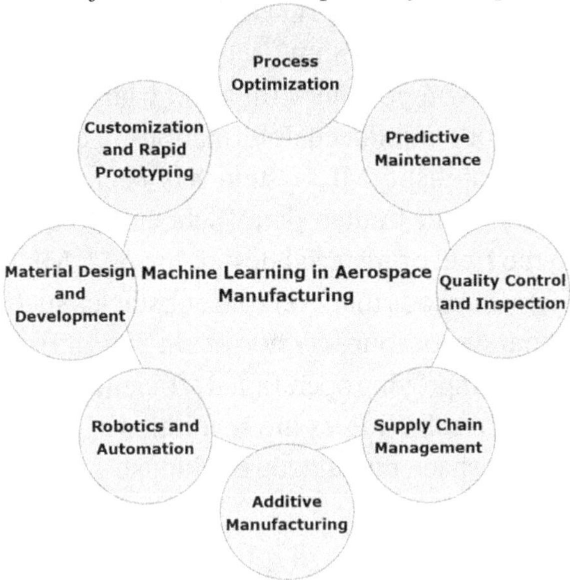

Process Optimization

For aerospace manufacturing, machining, welding and AM are processes demanding good control of many parameters. Historically, conditions have been optimum but a laborious process relying on trial and error. While that is possible, it is not perfectly efficient, ML can optimize these processes by analysing vast amounts of data from previous runs.

Indeed, ML algorithms can learn from historical data, and can identify the best parameter combination to achieve the best results for some tasks, e.g. determining the best temperature, pressure or speed for welding or machining. Further, by ensuring a consistent product quality, ML reduces material waste, reduces energy consumption and production time. In aerospace, with its extremely narrow tolerance for error, it is particularly important to optimize this.

By increasing operating efficiency, lowering costs, and improving product quality, ML plays a critical role in optimizing aerospace manufacturing processes. The application of ML techniques to several stages of the manufacturing process, from design to inspection, has demonstrated significant promise in addressing complex problems. The parameters of laser directed energy deposition (L-DED) have been successfully adjusted with the use of ML algorithms, which has allowed for the quicker prototyping of aircraft components with fewer trials (Ertugrul, 2024). To predict and reduce process-induced deformations in composite materials, a theory-driven probabilistic ML system has been developed, demonstrating effectiveness with limited data (Schoenholz & Zobeiry, 2023). When applied to the field of material design for AM, ML encourages the development of new methods that overcome obstacles such high costs and complex experimental procedures (Zhou et al., 2024). By automating the detection of flaws and improving operational efficiency through the analysis of real-time data, ML technologies are revolutionizing visual inspection techniques in the aerospace manufacturing industry (Rosell et al., 2023).

Predictive Maintenance

Manufacturing within aerospace involves high reliability and minimum downtime on complex machinery and equipment. Unexpected failure of equipment can delay and increase costs considerable. Predictive maintenance made possible with ML involves analysing sensor data generated from manufacturing equipment to predict failures before they take place with an algorithm.

It's because ML can alert maintenance teams to issues that they need to address, by identifying patterns and irregularities in real time, which can help timely interventions preventing breakdowns. Not only does this lengthen the lifespan of the machinery, but with this approach production schedules are continued which saves money due to costly downtime. In aerospace manufacturing, ML powered Predictive Maintenance has become standard practice, helping to make operations more efficient and reliable.

ML assumes a pivotal function in predictive maintenance (PdM) within the domain of aerospace manufacturing, thereby enhancing operational efficiency and minimizing expenditures. Through the utilization of sophisticated algorithms, manufacturers are able to anticipate equipment malfunctions and refine maintenance timelines, resulting in augmented operational dependability. Methodologies such as Convolutional Neural Networks (CNNs) and Long Short-Term Memory (LSTM) networks have demonstrated considerable potential in forecasting Remaining Useful Life (RUL) and State of Health (SOH) of machinery, attaining significant enhancements in accuracy (Habib & Mohamed, 2024).

The best predictive maintenance plan is based on operational effectiveness and product quality, which leads to a 50% decrease in machine downtime costs and a 64% decrease in scrap rates. The suggested framework has been empirically verified in an electromechanical component production line, resulting in a significant decrease in total costs (Riccio et al., 2024). The reliability of aircraft engines is greatly increased by the use of deep learning techniques in predictive maintenance. The aerospace manufacturing industry benefits greatly from this strategy, which uses data to enable continuous real-time monitoring, effectively reducing the risk of failures and optimizing maintenance schedules (Dube, 2024). The predictive maintenance paradigm in aerospace manufacturing relies heavily on machine learning, which makes use of sophisticated algorithms like auto encoders, artificial neural networks (ANN), and convolutional neural networks (CNN) to forecast maintenance needs and the remaining useful life of components (Amune et al., 2024).

Quality Control and Inspection

Aerospace manufacturing is critically dependent on the quality control of components. Quality control is assisted by ML to automate inspection process as well as to further improve defect detection accuracy. However, traditional inspection methods can be time consuming and often carry with them a high risk for human error when we are dealing with complex parts or large volumes.

Specialized ML driven systems, which involve using ML to analyse images, sensor data and other inputs to detect defects in manufactured components with great precision, possess the advantage of being cost effective. ML models can be trained to identify surface imperfections, dimensional deviations or material inconsistencies that are not visible to the naked eye. While the process of inspection does not change, ML is able to automate this process, significantly speeding up inspections and eliminating the possibility of components under inspection being approved for use which do not meet the highest standards.

In the field of aerospace manufacturing, ML plays a crucial role, particularly in relation to quality control and inspection procedures. By utilizing advanced algorithms and machine vision technologies, ML greatly improves the effectiveness, accuracy, and consistency of inspection procedures, which in turn leads to improved product quality. Machine vision technologies are frequently used for a wide range of inspection tasks, including assessments of adhesive application quality, assembly congruence, drilling precision, and surface integrity. These technologies make it easier to automate the measurement and identification of defects, which significantly reduces labour costs and lowers the hazards associated with human inspection techniques (Deng et al., 2024).

ML techniques, namely those applied to radiography analysis, enable the automated detection of flaws in aerospace welds, hence boosting the capacities of statistical analysis and improving the outcomes of quality control. It has been shown that a multi-model approach within this framework can lower false positive rates and increase inference velocity, allowing for real-time quality assessments (Tyystjärvi et al., 2024). Aircraft fuselage structural defects are detected using convolutional neural networks (CNNs), which leverage large annotated datasets to obtain detailed damage characteristics. Techniques like data augmentation and transfer learning greatly increase the robustness of these models, ensuring high accuracy in fault detection (Merola, 2024).

Supply Chain Management

The aerospace supply chain is highly complex, involving multiple suppliers, tight deadlines, and stringent quality requirements. ML is increasingly being used to optimize supply chain management by improving demand forecasting, inventory management, and logistics.

ML models can analyse historical data and external factors (such as market trends and geopolitical events) to predict future demand for components and materials. This allows manufacturers to better plan their inventory levels, reducing the risk of overstocking or stockouts. Additionally, ML can optimize logistics by predicting potential disruptions and suggesting alternative routes or suppliers. By enhancing supply chain resilience and efficiency, ML helps aerospace manufacturers maintain smooth production flows and meet delivery deadlines.

ML is revolutionizing operational effectiveness and decision-making frameworks in the aerospace manufacturing-supply chain management area. By employing advanced algorithms, companies can improve multiple aspects of their supply chains, including inventory management and preventive maintenance. ML algorithms improve manufacturing process planning and optimization, leading to increased production efficiency (Ilie & Semenescu, 2024). Predictive maintenance benefits greatly from ML, which helps anticipate equipment failures and minimize supply chain interruptions (Selvaraj & Lakshmanan, 2023). Throughout operational tasks, performance and dependability are strengthened by this anticipatory methodology. The capacity to examine large datasets enables firms to extract meaningful insights, which in turn improves decision-making effectiveness and fosters flexibility in responding to market demands (Sodiya et al., 2024).

Additive Manufacturing

AM technique, commonly referred to as 3D printing, is gaining prominence in the aerospace sector for the fabrication of intricate, lightweight components. ML is integral to the optimization of AM methodologies, facilitating the production of components that exhibit the requisite characteristics while minimizing defects.

By carefully examining variables including material composition, printing speed, and layer thickness, ML algorithms may predict the outcomes of the AM process. This feature makes it possible to make changes to the manufacturing process in real time, which guarantees better part quality and reduces the need for additional processing. Furthermore, ML can be used to create novel materials and structures tailored for AM, broadening the scope of what can be achieved with this cutting-edge technology.

AM technology in aerospace is undergoing a revolution in the fabrication of complex components with the integration of ML, which also improves material properties and operational efficiency. Particularly in the synthesis of functionally graded materials (FGMs), ML approaches have a critical role in addressing issues including process optimization, defect diagnosis, and real-time surveillance (Boopathy et al., 2024) (Karimzadeh et al., 2024). By analysing large and intricate datasets, ML makes it possible to create advanced materials with improved microstructural properties and performance measures (Ng et al., 2024). Physics-based ML frameworks combine data-centric approaches with physical principles to enhance the accuracy and clarity of manufacturing processes (Sharma et al., 2024).

Robotics and Automation

The integration of robotics and automation in the aerospace manufacturing industry is greatly boosted by machine learning, which enables robots to perform complex jobs with more precision and flexibility. Algorithms for ML enable robots to absorb data from their environment and gradually improve their operating efficiency.

For example, in tasks like applying coatings or assembling intricate parts, machine learning-powered robots demonstrate the ability to adapt to changes in the workplace and maintain consistent quality. Moreover, cobots (collaborative robots) with ML capabilities can work alongside human workers, learning from their actions and providing support for tasks requiring a high degree of accuracy or dexterity. The result of this collaboration between human operators and robotic systems is more flexible and efficient industrial processes.

The application of ML in the aerospace manufacturing domain, particularly in the areas of robotics and automation, is transforming the precision and efficiency of operations. Robotic systems can now perform complex tasks with greater accuracy and flexibility due to ML. This facilitates the optimization of efficient trajectories for robotic arms, significantly enhancing automation in manufacturing processes (Nahavandi et al., 2024). Advanced robotic systems are capable of adapting to new tasks and surroundings, reducing the need for substantial reprogramming (Singh et al., 2024). Artificial neural networks (ANNs) and support vector regression (SVR) are two ML techniques that have shown promise in reducing robotic errors, leading to notable decreases in mean error on many axes (McGarry et al., 2023).

Material Design and Development

In the field of aerospace manufacturing, the process of material selection is crucial since the materials that are selected must meet strict requirements for strength, weight, and heat resistance. Through the predictive study of their properties and performance, ML technologies are increasingly being integrated to speed up the design and development of innovative materials.

ML algorithms are able to analyse large datasets on material compositions and experimental test results, which makes it easier to find viable candidates for certain uses. By significantly reducing the time and cost associated with empirical testing, this approach allows aerospace companies to get innovative materials into the market more quickly. Additionally, by suggesting changes that improve the performance of current materials in specific aerospace applications, ML can improve their optimization.

The field of aerospace manufacturing is radically changing as a result of ML, particularly in the area of material development and design. Making use of large-scale datasets generated throughout the course of production processes, ML greatly increases the accuracy and speed of material selection and optimization procedures. ML techniques facilitate a deep understanding of the complex behaviours exhibited by fibre composite materials (FCMs), which are critical for aeronautical applications. Deep neural networks are among the techniques that can be used to represent the intricate interactions between material properties, improving performance evaluation and fault identification (Liu et al., 2024). ML techniques are additionally being applied to optimize the full material-process-structure-property continuum, thereby ensuring that the designs of materials are compatible with the capabilities of production (Morand et al., 2023).

Customization and Rapid Prototyping

The aircraft industry is increasingly moving toward customisation, where parts are carefully developed to meet exacting requirements. This evolution is supported by machine learning, which makes it easier to quickly generate prototypes and customize components.

By applying generative design, a machine learning-driven process, manufacturers can automatically generate a large number of design iterations based on specified characteristics like as weight, strength, and aerodynamic efficiency. This feature encourages the rapid development of prototypes that meet exacting requirements. Furthermore, without requiring a significant amount of machinery reconfiguration, ML can be used to adapt production processes in real-time to changes in design or material specifications.

Aerospace production is undergoing a fundamental transformation associated with ML, which is improving customization capabilities and speeding up the fast-prototyping processes. Its uses include manufacturing parameter optimization and inspection process automation, which significantly increase productivity and product quality. An example of rapid prototyping is the development of a Silicon Carbide (SiC) inverter for aerospace applications, which simplifies the design and testing stages through the use of simulations and hardware-in-the-loop testing techniques (Farnell et al., 2023) (Boopathy et al., 2022). Prototype design can be quickly iterated and modified due to ML and the integration of complex control systems (Brunton et al., 2021).

CHALLENGES AND LIMITATIONS

The challenges and limitations that accompany any cutting-edge technology also apply to machine learning in aeronautical design and manufacture. Before ML can be fully integrated into aircraft systems, it needs to overcome particular challenges, as shown in Figure 4, despite its revolutionary assurance.

Figure 4. Overcoming ML challenges in aerospace

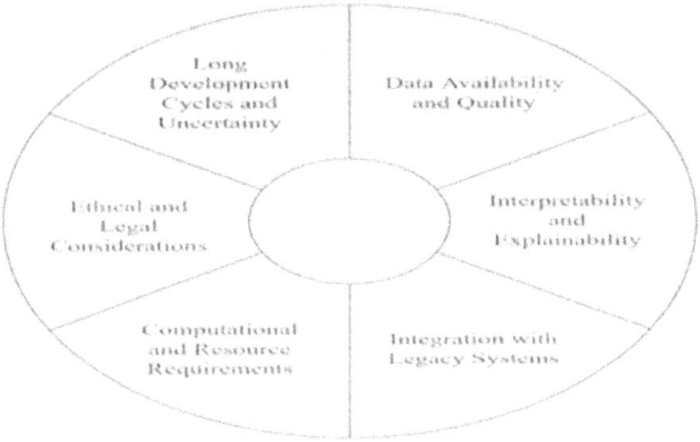

Data Availability and Quality

ML relies heavily on large datasets to train its algorithms. However, in aerospace, obtaining sufficient, high-quality data is a significant challenge for several reasons:

Limited Historical Data

Due to the high level of specialization that characterizes aerospace systems, there is a lack of publicly available data, especially for specialized applications. Due to competitive and security concerns, many aerospace companies carefully protect their private data, which limits data sharing throughout the industry.

High Cost of Data Collection

Unlike certain industries where data collection may be relatively simple, aerospace design, testing, and manufacture sometimes need expensive simulations, wind tunnel tests, and flying operations to obtain data. There may be insufficient datasets because the costs involved in acquiring such data can be unreasonable.

Data Inconsistency

Since aerospace projects take a long time to complete, the technology used to collect data may evolve over time. When fed into ML algorithms, inconsistent data from several sources or time periods might result in flawed models or inaccurate predictions.

Class Imbalance

It can be challenging to train models to precisely forecast uncommon but crucial failure modes because essential aerospace systems typically have a lot more information on normal operation than on failure events. For instance, while airplane engines are often quite dependable, there is little failure data available.

Interpretability and Explainability

Interpretability of the models is one of the major challenges in the field of machine learning, especially in mission-critical industries like aerospace:

Black-Box Nature

The Black-Box Nature: Many ML algorithms, particularly deep learning algorithms, are "black boxes." Even while these models might produce accurate forecasts, it might be challenging to understand the reasoning behind the model's choices. Understanding how and why a model arrived at a given conclusion is crucial for engineers and regulatory agencies working in the aerospace sector, where safety is of the utmost concern, especially in situations involving human life.

Regulatory Scrutiny

ML models must be explainable due to the strict regulations that regulate the aerospace industry, particularly those enforced by agencies like the European Aviation Safety Agency (EASA) and the Federal Aviation

Administration (FAA). To ensure that these models adhere to safety and reliability standards, aerospace engineers and regulatory bodies need to be able to outline the decision-making processes of these models.

Risk of Overfitting

ML models have a tendency to "overfit" to specific data patterns, which means that while they may perform well on training datasets, they may not be able to generalize well to new, unseen data. In the aerospace industry, where key system failures could arise from an inability to accurately predict real-world operating performance, such a failure could prove tragic.

Integration with Legacy Systems

The aerospace industry is extremely reluctant to adopt new technology because of the strict safety, testing, and regulatory requirements that control its activities. A lot of the current aerospace systems have been in use for decades, and there are significant challenges in integrating ML technologies into these outdated systems:

Compatibility Issues

Outdated hardware or software architectures are commonly used in aerospace design and manufacturing frameworks. The implementation of modern ML algorithms on these old systems requires extensive system upgrades or customisation, which can be costly and time-consuming.

Resistance to Change

The safety-critical nature of the aerospace industry makes it inherently conservative. When it comes to replacing or improving existing systems with relatively new ML solutions that carry unexpected risks, engineers and management people may be reluctant. It can be hard to convince stakeholders of the reliability and benefits of machine learning, especially when the costs and risks of integration are high.

Cyber Security Risks

Aerospace systems are becoming more endangered by cyber-attacks as ML and data-driven decision-making become more integrated into their operations. An emerging challenge for aircraft designers is protecting ML models from manipulation or hostile attacks, in which criminals may purposefully alter input data to produce inaccurate predictions.

Computational and Resource Requirements

Although ML has advanced significantly, its broad use in aerospace may be constrained by its computational and resource-intensive requirements:

High Computational Costs

Extensive computing resources, specialized hardware configurations like Graphics Processing Units (GPUs), and huge data storage capacities are required for the training of ML models, particularly those based on deep learning architectures. In the aerospace industry, which usually involves complex, high-dimensional datasets, model simulation and training can be quite costly, especially for smaller organizations.

Real-Time Performance Requirements

Some applications in the aircraft industry require the ability to make decisions instantly (e.g., autonomous flying systems and predictive maintenance protocols). It's possible that ML models will not digest data quickly enough to meet the demanding real-time operational requirements of aircraft systems. This limitation may make it more difficult to implement ML in vital area like control systems.

Energy Efficiency

There are energy and computational resource constraints for many aerospace applications, including on board systems in aircraft and spacecraft. The implementation of complex ML algorithms in these settings might need power consumption that is higher than the available capacity, which would make it more difficult to use ML effectively in these situations.

Ethical and Legal Considerations

Significant moral and legal challenges are brought about by the aerospace industry's growing reliance on machine learning, especially in relation to accountability and decision-making:

Accountability in Case of Failure: It is unclear who should be held accountable when a ML model makes a significant mistake, like making incorrect forecasts about aircraft component failures or autonomous flight decisions. Does the developer of the algorithm, the manufacturer that implemented it or the operator that used it hold the responsibility? The integration of ML in the aerospace industry is greatly impeded by this unclear area of accountability.

Bias and Fairness: It is possible for ML methods to deliberately reinforce biases originating from the datasets they are trained on. If a ML model is used in safety-critical areas, including predicting pilot conduct or assessing flight operations hazards, and these predictions are biased, it could lead to unfair or dangerous decisions. In a globally networked business such as aerospace, where activities cover numerous geographic, social, and cultural frameworks, the assurance of equity and the elimination of bias from ML models are critical.

Certification and Approval: Due to strict regulatory monitoring in the aerospace industry, any new technology including ML must go through a rigorous certification process. Because ML relies on constantly updated datasets and its decision-making processes are flexible, the current certification frameworks might not be sufficient to handle its complexities. Because of this, regulatory agencies need to develop new guidelines for assessing, certifying, and approving ML systems in the aerospace industry, which could ultimately slow the rapid uptake of this cutting-edge technology.

Long Development Cycles and Uncertainty

Projects in the aerospace industry typically have lengthy development schedules that span several years or even decades. This time frame offers unique difficulties for integrating rapidly evolving technology like machine learning:

Pace of Technological Change

The field of ML is known for its rapid growth, with new techniques, tools, and algorithms appearing on a regular basis. The slow pace of aerospace development suggests that a new ML technique may already be outdate or replaced by a more efficient substitute by the time it is fully integrated into a design or production process.

Uncertainty in Outcomes

ML adds a level of uncertainty, in contrast to traditional predictive engineering approaches that enable accurate outcome predictions. Because ML systems have no assurance that they will produce precise, reliable outcomes, engineers and designers may be reluctant to put their trust in them. The inventive potential of ML and the inherent risks of uncertainty in critical systems must be balanced by aerospace manufacturers.

FUTURE TRENDS AND OPPORTUNITIES

The aircraft sector could undergo a significant transformation as ML develops further. The future is full with exciting possibilities, from enabling new technologies like autonomous systems and smart maintenance to increasing the efficiency of design and production processes. In this section, Figure 5 surveys the primary future advancements and opportunities for ML in aerospace.

Figure 5. Future trends of ML in aerospace

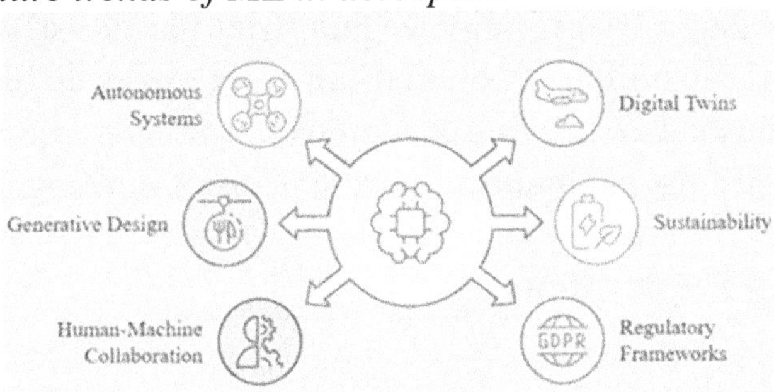

Advanced Autonomous Systems and AI-Driven Flight

The development of more complex autonomous systems and AI-enhanced aviation technology is expected to heavily rely on machine learning. ML will likely play a key role in autonomous aircraft, UAVs, and systems intended for space travel in the future. Significant advancements are anticipated in a number of areas, including:

Autonomous Aircraft and UAVs

Autonomous aircraft and UAVs are expected to become more popular for military, commercial, and exploration applications. These systems will be able to do complex tasks including route optimization, collision avoidance, weather prediction, and mission planning with the help of ML algorithms. It might soon be possible for airplanes to fly completely autonomously, managing landing, take-off, and in-flight adjustments.

Swarm Intelligence

The development of swarm technologies in which several autonomous drones or airplanes work together will be accelerated by ML. This innovation has the potential to revolutionize industries like disaster relief, cargo transportation, and military operations. In order to successfully complete tasks while adjusting to changing conditions and mission requirements, ML will enable real-time communication and decision-making among units.

AI for Space Exploration

By enabling the development of increasingly self-sufficient and intelligent spacecraft, ML has the potential to drastically change space travel. Future journeys to the Moon, Mars, and other planets will require artificial intelligence systems that can process large amounts of sensor data, navigate unfamiliar situations, and make choices in real time without human intervention.

Digital Twins and Predictive Maintenance

The aerospace industry could undergo a change because to the concept of digital twins, which are virtual replicas of physical assets. This change will be facilitated by the incorporation of ML, offering so far unusual opportunities in lifecycle management, real-time supervision, and predictive maintenance.

Digital Twin Technology

A digital twin is a computerized model of a physical system, like an engine or airplane, which precisely replicates its real-time operating performance. By continuously analyzing data from the real system and its virtual equivalent, ML will support digital twin frameworks by predicting potential problems, improving operational efficiency, and extending component longevity. This feature will enable aerospace operators and manufacturers to simulate different situations, detect degradation, and fix problems before they appear.

Predictive Maintenance

Upcoming ML models will show greater competence in precisely predicting component failures and maintenance needs prior to issues developing. ML can forecast the probability of component failures and suggest preventive maintenance procedures by analyzing large datasets from aircraft sensors and historical records. This strategy will reduce unexpected downtime, save costs, and improve the fleet's overall dependability.

Lifecycle Management

Digital twins with ML capabilities will also be crucial for managing the entire aircraft component lifecycle, which includes design, production, operational use, and final disposal. Organizations can achieve increased sustainability and efficiency by optimizing each stage of this lifecycle.

Generative Design and AM

Though generative design and AM technology are already being investigated in the aerospace industry, ML is expected to propel these technologies to levels never before seen in the years to come.

Generative Design

Based on input restrictions like as material qualities, weight, and strength, ML algorithms produce a variety of design alternatives in generative design. Generative design tools will become increasingly smart as ML advances, enabling them to generate innovative designs that surpass those developed by humans. ML-designed aircraft components of the future will be tailored for particular purposes, lowering weight and material consumption without sacrificing structural integrity.

Advanced AM

In the aerospace industry, AM is expanding quickly, but ML has the potential to further enhance its potential. By predicting how different materials would behave during the printing process, future ML models will improve AM procedures, guaranteeing improved quality control and reducing errors. In order to improve material qualities and speed up production cycles, ML will also be used to monitor and alter the printing process in real-time.

Topology Optimization

More complex topology optimization techniques, in which component composition and geometry are adjusted for maximum strength and weight efficiency, will be made possible by ML techniques. The most effective material distributions for components will be determined by ML models, producing structures that are stronger and lighter than traditional design, these two qualities that are crucial for aeronautical application.

Sustainability and Green Aviation

Sustainability has become a top priority for the future of the aircraft industry due to the growing need to reduce pollution. By improving fuel efficiency, reducing emissions, and enabling the development of alternative propulsion technologies, ML is expected to play a key role in achieving environmentally friendly aviation operations.

Fuel Efficiency Optimization

In order to improve fuel efficiency and reduce emissions, ML algorithms can be used to optimize engine operations and fly paths in real-time. Artificial intelligence systems, for example, would carefully evaluate weather patterns, environmental factors, and aircraft performance indicators to suggest the best routes and altitudes, reducing fuel use.

Electric and Hybrid Propulsion

In an effort to reduce carbon emissions, the aircraft industry is looking at the use of electric and hybrid-electric propulsion systems. In order to optimize battery usage, power management, and propulsion efficiency, MLis expected to be essential to the development and implementation of these cutting-edge propulsion systems. ML will be essential to making electric airplanes and unmanned aerial vehicles financially feasible as these technologies become more popular.

Material Innovation for Sustainability

Additionally, ML will help manufacture aircraft using stronger, lighter, and more environmentally sound materials. ML, by analyzing large datasets about material properties, can help find new materials that offer the best combination of durability, weight reduction, and environmental sustainability, ultimately leading to a more sustainable aerospace manufacturing process.

Enhanced Human-Machine Collaboration

The line separating autonomous machine operations from human cognitive processes will become increasingly blurred as ML advances. In the future, human operators and AI systems will work in greater sync in the aircraft industry, leading to better decision-making, more robust safety protocols, and more operational effectiveness.

Human-AI Teaming

Within the cockpit, ML will help pilots make better decisions by providing real-time data analytics, suggesting the best flying paths, and carrying out thorough risk assessments. It is anticipated that this expanded collaboration will increase overall safety and reduce the rate of human mistake. Additionally, in certain situations, artificial intelligence will act as a co-pilot, with the capacity to take on certain duties in challenging or stressful situations.

AI-Driven Decision Support

In order to provide aerospace engineers and operators with advanced decision support mechanisms, advanced ML frameworks will be employed. For example, ML will speed up and improve the decision-making process during the design phase by offering analytical insights and suggestions based on previous designs. Similarly, AI-driven solutions will assist manufacturing workers by identifying possible flaws or instantly streamlining assembly processes.

Augmented Reality (AR) and Virtual Reality (VR)

In the near future, AR and VR will also be combined with ML technology to create immersive training and design environments for aviators, aerospace engineers, and designers. These cutting-edge tools will enable artificial intelligence-powered interactive simulations, giving people the opportunity to interact with complex systems in a safe and controlled setting.

Regulatory Frameworks and Standardization

In order to ensure safety, dependability, and compliance, new regulatory frameworks and standards will need to be developed as ML is increasingly incorporated into aircraft systems. In the future, it is expected that aerospace companies, authorities, and standards groups will work together more closely to develop policies that govern the use of ML in key areas.

Certification of AI Systems

The official certification of AI and ML systems designed for safety-critical tasks will be a major trend. For the evaluation, verification, and certification of AI systems used in aerospace design, manufacturing, and operations, regulatory bodies such as the Federal Aviation Administration

(FAA) and the European Aviation Safety Agency (EASA) must set clear guidelines and procedures.

Ethics and Accountability

There will probably be a greater focus on developing ethical frameworks and accountability systems as ML takes on steadily more important decision-making responsibilities in the aerospace industry. Who is accountable in the event that an artificial intelligence system fails? What steps may be taken to ensure that AI systems make fair and impartial decisions? These questions need to be carefully considered as essential elements of the future regulatory structure.

Global Standards for Data and AI Integration

The development of global guidelines for data gathering, sharing, and AI model integration will be extremely beneficial to the aerospace industry. By facilitating improved cooperation between global aerospace companies, regulatory agencies, and governmental organizations, the standardization process will ensure that ML technologies can be successfully applied across national borders while abiding by generally recognized safety and operational standards.

REFERENCES

Action, J., Morehead, K., Hiromoto, T., & Gajus, B. (2024). Application of Generative Design to Thin-Walled Aerostructure. In AIAA SCITECH 2024 Forum (p. 0362). DOI: 10.2514/6.2024-0362

Adimurthy, V., Balu, R., Priyadarshi, P., Ramanan, R. V., & Ravikumar, C. (2005). Multidisciplinary Optimization Aerospace Design: The Emerging Technology For Complex Systems. *Journal of Aerospace Sciences and Technologies*, •••, 79–92.

Amune, A., Shahari, S., Kasurde, S., Nimale, S., Surdas, S., & Tayde, S. (2024, April). Exploring Predictive Maintenance and Signal Processing Techniques for Automotive Health Monitoring. In *2024 International Conference on Expert Clouds and Applications (ICOECA)* (pp. 541-547). IEEE. DOI: 10.1109/ICOECA62351.2024.00100

Anjana, P. K., Abhilash, P. V., Bisariya, S., & Sutrakar, V. K. (2024). Inverse Approach for Metasurface Based Radar Absorbing Structure Design for Aerospace Applications Using Machine Learning Techniques (No. 2024-26-0480). SAE Technical Paper.

Bidar, O., He, P., Anderson, S., & Qin, N. (2024). Aerodynamic shape optimisation using a machine learning-augmented turbulence model. In *AIAA SCITECH 2024 Forum* (p. 1231). DOI: 10.2514/6.2024-1231

Boopathy, G., Gurusami, K., Chinnapandian, M., & Vijayakumar, K. R. (2022). Optimization of Process Parameters for Injection Moulding of Nylon6/SiC and Nylon6/B_4C Polymer Matrix Composites. *Fluid Dynamics & Materials Processing*, *18*(2), 223–232. DOI: 10.32604/fdmp.2022.018225

Boopathy, G., Palani, S., Ganesan, B., & Kumaar, K. (2024). Advancements in Design, Fabrication, and Applications of Functionally Graded Composite Materials. In *Engineering Materials for Efficient Energy Storage and Conversion* (pp. 25–50). IGI Global. DOI: 10.4018/979-8-3693-2798-2.ch002

Borwankar, P., Kapania, R. K., Inoyama, D., & Stoumbos, T. (2024). Multidisciplinary Design Analysis and Optimization of Space Vehicle Structures. In *AIAA SCITECH 2024 Forum* (p. 2898). DOI: 10.2514/6.2024-2898

Boy, G. A., & Morel, C. (2022). The machine as a partner: Human-machine teaming design using the PRODEC method. *Work (Reading, Mass.)*, *73*(s1), S15–S30. DOI: 10.3233/WOR-220268 PMID: 36214030

Brunton, S. L., Nathan Kutz, J., Manohar, K., Aravkin, A. Y., Morgansen, K., Klemisch, J., & McDonald, D. (2021). Data-driven aerospace engineering: Reframing the industry with machine learning. *AIAA Journal*, *59*(8), 2820–2847. DOI: 10.2514/1.J060131

Caliari, T., & Ferreira, M. J. B. (2023). The historical evolution of the Brazilian aeronautical sector: A combined approach based on mission-oriented innovation policy (MOIP) and sectoral innovation system (SIS). *Economics of Innovation and New Technology*, *32*(5), 682–699. DOI: 10.1080/10438599.2021.2011258

Carlo, R. (2024, June 27). Marialuisa, Menanno., Ilenia, Zennaro., Matteo, Mario, Savino. (2024). 3. A New Methodological Framework for Optimizing Predictive Maintenance Using Machine Learning Combined with Product Quality Parameters. *Machines (Basel)*, *12*(7), 443. Advance online publication. DOI: 10.3390/machines12070443

Deng, L., Liu, G., & Zhang, Y. (2024, April). A Review of Machine Vision Applications in Aerospace Manufacturing Quality Inspection. In 2024 4th International Conference on Computer, Control and Robotics (ICCCR) (pp. 31-39). IEEE. DOI: 10.1109/ICCCR61138.2024.10585378

Du, X., Martins, J. R., O'Leary-Roseberry, T., Chaudhuri, A., Ghattas, O., & Willcox, K. E. (2023). Learning optimal aerodynamic designs through multi-fidelity reduced-dimensional neural networks. In AIAA SCITECH 2023 Forum (p. 0334). DOI: 10.2514/6.2023-0334

Dube, A. (2024). Application of Deep Learning in Predictive Maintenance of Aircraft Engines. *Darpan International Research Analysis*, *12*(3), 83–100. DOI: 10.36676/dira.v12.i3.58

Ertugrul, G., Alimov, A., Sviridov, A., & Härtel, S. Machine learning application for optimization of laser directed energy deposition process for aerospace component rapid prototyping in additive manufacturing. Materials Research Proceedings, 41.

Eskandariyun, A., Fu, H., Stere, A., Bauer, A., Reynolds, C., Chengalva, M., . . . Zobeiry, N. (2024, April). A Novel Machine Learning Framework for Digital Estimation of Allowables in Aerospace Composites. In ASME Aerospace Structures, Structural Dynamics, and Materials Conference (Vol. 87745, p. V001T01A014). American Society of Mechanical Engineers. DOI: 10.1115/SSDM2024-121597

Farnell, C., Jackson, J., Corbitt, A., & Mantooth, H. A. (2023, September). Rapid prototyping of a SiC-based PMSM motor drive for aerospace applications. In *2023 IEEE Design Methodologies Conference (DMC)* (pp. 1-5). IEEE. DOI: 10.1109/DMC58182.2023.10412479

Gao, Y. (2023). CFD based aerodynamic optimization design. *Theoretical and Natural Science, 18*(1), 83–90. DOI: 10.54254/2753-8818/18/20230343

Gómez-Carmona, O., Casado-Mansilla, D., López-de-Ipiña, D., & García-Zubia, J. (2024). Human-in-the-loop machine learning: Reconceptualizing the role of the user in interactive approaches. *Internet of Things : Engineering Cyber Physical Human Systems, 25*, 101048. DOI: 10.1016/j.iot.2023.101048

Govindarajan, B., Jaganraj, R., Srinivasan, V., & Kumaran, T. (2025). Evolution of UAV Technology From Early Innovations to Future Horizons. In *Innovations and Developments in Unmanned Aerial Vehicles* (pp. 53–94). IGI Global Scientific Publishing.

Habib, M. K., & Mohamed, K. (2024, August). Enhancing Predictive Maintenance Hyperparameter Optimization and Adopted Strategies. In *2024 IEEE International Conference on Mechatronics and Automation (ICMA)* (pp. 153-158). IEEE. DOI: 10.1109/ICMA61710.2024.10633049

Hajela, P., Sakalkar, V., & Mullur, A. (2009). Multidisciplinary Analysis and Design Tools For Uncertainty Modeling. *Journal of Aerospace Sciences and Technologies*, •••, 240–251.

Humml, J. M., Oshima, E., O'Gara, S., Rusch, A., Gharib, M., Lee, V., & Khodadoust, A. (2024). Development of machine learning tools for aerospace design: wind tunnel investigations on a speed bump model. In AIAA SCITECH 2024 Forum (p. 2834). Ilie, M., & Semenescu, A. Supply Chain Management Of Manufacturing Processes Using Machine Learning Technique. DOI: 10.2514/6.2024-2834

Jose, L. A.Jr, Brintrup, A., & Salonitis, K. (2020). Analysing the evolution of aerospace ecosystem development. *PLoS One*, *15*(4), e0231985. DOI: 10.1371/journal.pone.0231985 PMID: 32343729

Kalyan, B. S. (2024, February). A multi-fidelity platform using machine learning for integrated design of aircraft systems. In *2024 International Conference on Emerging Systems and Intelligent Computing (ESIC)* (pp. 243-248). IEEE. DOI: 10.1109/ESIC60604.2024.10481666

Karimzadeh, M., Basvoju, D., Vakanski, A., Charit, I., Xu, F., & Zhang, X. (2024). Machine Learning for Additive Manufacturing of Functionally Graded Materials. *Materials (Basel)*, *17*(15), 3673. DOI: 10.3390/ma17153673 PMID: 39124337

McGarry, L., Butterfield, J., Murphy, A., & Higgins, C. (2023). Machine learning methods to improve the accuracy of industrial robots. SAE International Journal of Advances and Current Practices in Mobility, 5(2023-01-1000), 1900-1918.

Merola, S., Guida, M., & Marulo, F. Digital optics and machine learning algorithms for aircraft maintenance. Materials Research Proceedings, 42.

Mingyu, L. (2024, August). Haotian, Li., Hongyuan, Zhou., Hong, Zhang., Guangyan, Huang. (2024). 1. Development of machine learning methods for mechanical problems associated with fibre composite materials: A review. *Composites Communications*, *49*, 101988. Advance online publication. DOI: 10.1016/j.coco.2024.101988

Mishra, R., Ranjan Kumar, H., Nayak, S., Nayak, S. K., Mishra, S. B., & Nanda, B. K. (2022, November). Structural Analysis of Aircraft Wing Considering Titanium Alloy. In International conference on Advances in Materials and Manufacturing (pp. 171-178). Singapore: Springer Nature Singapore.

Morand, L., Iraki, T., Dornheim, J., Sandfeld, S., Link, N., & Helm, D. (2023). Machine learning for structure-guided materials and process design. arXiv preprint arXiv:2312.14552.

Muelaner, J. E. (2024). Generative Design in Aerospace and Automotive Structures (No. EPR2024016). SAE Technical Paper.

Mumtaz, A. Shah. (2015). 6. Commercial Aerospace Industry from the U.S. Monopoly to Duopoly. Social Science Research Network, doi: DOI: 10.2139/SSRN.2545501

Nahavandi, S., Alizadehsani, R., Nahavandi, D., Lim, C. P., Kelly, K., & Bello, F. (2024). Machine learning meets advanced robotic manipulation. *Information Fusion, 105*, 102221. DOI: 10.1016/j.inffus.2023.102221

Ng, W. L., Goh, G. L., Goh, G. D., Ten, J. S. J., & Yeong, W. Y. (2024). Progress and opportunities for machine learning in materials and processes of additive manufacturing. *Advanced Materials, 36*(34), 2310006. DOI: 10.1002/adma.202310006 PMID: 38456831

Paleja, R., Munje, M., Chang, K., Jensen, R., & Gombolay, M. (2024). Designs for Enabling Collaboration in Human-Machine Teaming via Interactive and Explainable Systems. arXiv preprint arXiv:2406.05003.

Pop, G. I., Titu, A. M., & Pop, A. B. (2023). Enhancing Aerospace Industry Efficiency and Sustainability: Process Integration and Quality Management in the Context of Industry 4.0. *Sustainability (Basel), 15*(23), 16206. DOI: 10.3390/su152316206

Raja, S., Al-Tmimi, H. M., Ghadir, G. K., Mustafa, M. A., Alani, Z. K., Rusho, M. A., & Rajeswari, N. (2024). An analysis of polymer material selection and design optimization to improve Structural Integrity in 3D printed aerospace components. Applied Chemical Engineering, 1875-1875.

Rosell, A., Svenman, E., Westphal, P., Mukundan, A., Bhattacharya, S., Bharthulwar, S., . . . Jhanardhanan, S. (2023, September). Machine learning-based system to automate visual inspection in aerospace engine manufacturing. In 2023 IEEE 28th International Conference on Emerging Technologies and Factory Automation (ETFA) (pp. 1-8). IEEE. DOI: 10.1109/ETFA54631.2023.10275515

Sarojini, D., & Mavris, D. (2022). Structural Analysis and Optimization of Wings Subjected to Dynamic Loads. *AIAA Journal, 60*(2), 1013–1023. DOI: 10.2514/1.J060931

Schoenholz, C., & Zobeiry, N. (2024). An accelerated process optimization method to minimize deformations in composites using theory-guided probabilistic machine learning. *Composites. Part A, Applied Science and Manufacturing, 176*, 107842. DOI: 10.1016/j.compositesa.2023.107842

Selvaraj, K., & Lakshmanan, S. (2021). The Machine learning for predictive maintenance in supply chain management. Journal of Artificial intelligence and Machine Learning, 1(1), 9-15.

Sharma, R., Raissi, M., & Guo, Y. B. (2024). Physics-Informed Machine Learning for Smart Additive Manufacturing. arXiv preprint arXiv:2407.10761.

Shirvani, A., Nili-Ahmadabadi, M., & Ha, M. Y. (2023). Machine learning-accelerated aerodynamic inverse design. *Engineering Applications of Computational Fluid Mechanics, 17*(1), 2237611. DOI: 10.1080/19942060.2023.2237611

Singh, M., & Khan, S. A. L. A. (2024). Advances in Autonomous Robotics: Integrating AI and Machine Learning for Enhanced Automation and Control in Industrial Applications. *International Journal for Multidimensional Research Perspectives, 2*(4), 74–90. DOI: 10.61877/ijmrp.v2i4.135

Sodiya, E. O., Jacks, B. S., Ugwuanyi, E. D., Adeyinka, M. A., Umoga, U. J., Daraojimba, A. I., & Lottu, O. A. (2024). Reviewing the role of AI and machine learning in supply chain analytics. *GSC Advanced Research and Reviews, 18*(2), 312–320. DOI: 10.30574/gscarr.2024.18.2.0069

Sunil, M. (2013). 8. Evolution of the sectoral system of innovation of India's aeronautical industry. *International Journal of Technology and Globalisation, 7*(1/2), 92. Advance online publication. DOI: 10.1504/IJTG.2013.052033

Tavares, S. M. O., Ribeiro, J. A., Ribeiro, B. A., & de Castro, P. M. S. T.Sérgio. (2024, March 22). Mo, Tavares., João, A., Ribeiro., Bruno, A., Ribeiro., Paulo, M.S.T., de, Castro. (2024). 4. Aircraft Structural Design and Life-Cycle Assessment through Digital Twins. *Designs*, *8*(2), 29. Advance online publication. DOI: 10.3390/designs8020029

Tong, M. T. (2023, June). Aero-Engines AI-A Machine-Learning App for Aircraft Engine Concepts Assessment. In Turbo Expo: Power for Land, Sea, and Air (Vol. 86939, p. V001T01A017). American Society of Mechanical Engineers. DOI: 10.1115/GT2023-102024

Tyystjärvi, T., Fridolf, P., Rosell, A., & Virkkunen, I. (2024). Deploying Machine Learning for Radiography of Aerospace Welds. *Journal of Nondestructive Evaluation*, *43*(1), 24. DOI: 10.1007/s10921-023-01041-w

Vignesh, P., Maheswari, R., Vijaya, P., & Vignesh, U. (2024). Machine Learning for Aerospace Object Categorization. In AI and Blockchain Optimization Techniques in Aerospace Engineering (pp. 164-180). IGI Global. DOI: 10.4018/979-8-3693-1491-3.ch008

Wang, W., & Ma, J. (2024). A review: Applications of machine learning and deep learning in aerospace engineering and aero-engine engineering. *Advances in Engineering Innovation*, *6*(1), 54–72. DOI: 10.54254/2977-3903/6/2024060

Zhizhong, C., Hongqing, L., Zengcong, L., Yan, C., Jie, C., & Xiaoqi, L. (2023). Structural lightweight design and experimental validation for aerospace sealed cabin. *Frontiers of Mechanical Engineering*, *9*, 1265734. DOI: 10.3389/fmech.2023.1265734

Zhou, H. R., Yang, H., Li, H. Q., Ma, Y. C., Yu, S., Shi, J., Cheng, J., Gao, P., Yu, B., Miao, Z., & Wei, Y. P. (2024). Advancements in machine learning for material design and process optimization in the field of additive manufacturing. *China Foundry*, *21*(2), 101–115. DOI: 10.1007/s41230-024-3145-3

Chapter 2
Machine Learning-Based Solutions for Aerospace Engineering

G. Prasad
https://orcid.org/0000-0002-5709-9182
Chandigarh University, Punjab, India

ABSTRACT

The incorporation of machine learning (ML) in aircraft engineering has transformed the design, analysis, and operation of intricate aerospace systems. This study examines the present and developing applications of machine learning techniques in critical domains like aircraft design optimisation, defect detection and diagnostics, flight control systems, and predictive maintenance. Utilising extensive information from simulations, sensors, and real-time operations, machine learning models facilitate more efficient decision-making, improved system reliability, and decreased operational costs. Moreover, progress in deep learning, reinforcement learning, and neural networks is being progressively utilised for applications spanning aerodynamic modelling to autonomous flight control. This study emphasises the difficulties related to data quality, interpretability, and model validation in safety-critical aircraft contexts.

DOI: 10.4018/979-8-3693-7525-9.ch002

INTRODUCTION

In an age characterised by technological advancement, the incorporation of Artificial Intelligence (AI) and Machine Learning (ML) has become a revolutionary influence across multiple sectors. The aerospace and defence sectors are leading in innovation by utilising AI and ML to transform the functionalities of contemporary aircraft, spacecraft, and military operations. The collaboration between advanced technology and intricate engineering has transformed conventional methods and created opportunities for innovative applications previously considered science fiction. The importance of AI and ML in aerospace and defence is paramount. The intricacies inherent in these fields, where safety, precision, and real-time decision-making are crucial, have prompted the development of AI-driven solutions. This blog examines the diverse uses of AI and ML in aerospace and defence, investigating their influence on the future of aviation, space exploration, and military strategy.

The aviation industry is evolving due to artificial intelligence (AI) and machine learning (ML), which enhance production while upholding stringent safety standards. Artificial Intelligence and Machine Learning are demonstrating their utility in enhancing customer experience and optimising operations, notwithstanding the aviation industry's traditionally conservative stance on adopting new technologies. AI-powered chatbots are now accessible around the clock to assist, addressing common enquiries from travellers concerning luggage policies, flight schedules, and other subjects. By automating these interactions, airlines can allocate human resources to more complex consumer requirements. Artificial Intelligence and Machine Learning are revolutionising aircraft maintenance as well as customer service. Predictive maintenance, facilitated by sophisticated data analytics, enables airlines to detect potential mechanical issues prior to their escalation. This proactive strategy ensures passenger safety, enhances operational efficiency, and minimises unforeseen downtime. Furthermore, artificial intelligence is essential for optimising in-flight operations. AI-driven systems can suggest fuel-efficient routes by analysing real-time data, so reducing carbon emissions and fuel consumption. AI-driven technologies are aiding the airline industry in minimising its environmental footprint while maintaining economic efficiency as sustainability gains traction.

BACKGROUND OF THE RESEARCH

AI and ML are transforming the aerospace manufacturing sector by proactively addressing maintenance issues, extending aircraft operational duration, and reducing expenses. They facilitate enhanced flight planning, hence diminishing emissions and fuel consumption. These technologies enable cockpit automation, optimised supply chains, enhanced quality control, and improved passenger experiences. This section discusses the application of AI and ML in the aerospace sector.

Barkalov et al. (2022). Machine learning in conjunction with the worldwide search method is employed to optimise the parameters of the turbulence model. Beetham and Capecelatro (2020) developed turbulence closures through sparse regression with incorporated form invariance. Bounds et al. (2023) The Optimisation of Turbulence Modelling Closure The coefficients Using an Explainability-Based Machine Learning Algorithm. Brenner et al. (2019) A perspective on the application of machine learning to the advancement of fluid mechanics. Brunton and Kutz (2022) Dynamical systems, control, and machine learning are all examples of data-driven science and engineering.

Dangeti (2018) The Dynamic Forecasts of Flow over a Reduced Road Vehicle: The Importance of Turbulence Closure Coefficients. Eivazi et al. (2022) In the direction of the extraction of orthogonal and parsimonious non-linear modes from turbulent flows. Godunov and Bohachevsky (1959) developed the finite difference method for the numerical computation of irregular solutions to the equations of fluid dynamics. Guilmineau et al. (2018) Evaluate hybrid RANS-LES formulations for flow simulation in the vicinity of the Ahmed body. Jeon et al. (2022) Finite volumetric method network for the speeding up of uncertain computational fluid dynamics: Non-reacting and reacting flows.

Kraichnan (1967) Inertial ranges in two-dimensional turbulence. Ling et al (2016) employed deep neural networks with incorporated invariance to model Reynolds averaged turbulence. An incremental machine-learning system for RANS turbulence modelling, Liu et al. (2021). Maulik et al. (2019) utilised neural networks to model subgrids for two-dimensional turbulence. Morita et al. (2022) Utilising Gaussian process regression in Bayesian optimisation to address computational fluid dynamics issues.

Obiols-Sales et al. (2020) CFDNet is a fluid simulation accelerator that is founded on deep learning. Pope (1999) A viewpoint on the modelling of turbulence. The correctness of RANS models with DNS data was the subject of Poroseva et al. (2016). Shan et al. (2020) A study on the development of a rapid solver for Poisson's equation using a deep learning technique. Shi et al. (2024) CFD-ML-driven structural optimisation and experimental analysis of thermal plasma reactors: From simulation to actuality. Spalart (2000) Strategies for the simulation and modelling of turbulence. Spalart and Allmaras (1992). A single-equation turbulent model for aerodynamic flows. Vinuesa et al. (2017) Pressure-gradient chaotic boundary layers are layers that develop around a wing section. Implementation of artificial intelligence in mathematical fluid dynamics by Wang and Wang (2021). Weller et al. (1998) employed object-oriented techniques to develop a tensorial technique for computation continuum mechanics. Wu et al. (2019) Reynolds-averaged conditioned Navier–Stokes equations may be characterised by explicit data-driven Reynolds stress closure. The estimation of turbulent closure factors for data centres using artificial intelligence algorithms was conducted by Yarlanki et al. (2012). In 2019, Zhu et al. proposed artificial intelligence methods for the modelling of turbulence in subsonic flows surrounding airfoils.

PREDICTIVE MAINTENANCE

Maintenance is essential for ensuring an aircraft's safety. An aircraft may encounter several unexpected maintenance issues. The aviation sector use AI-enabled predictive maintenance to address these issues. Predictive maintenance enables the rapid identification and reporting of potential issues in real-time. It ensures that the process plan is more efficient and expedited by forecasting the schedule for repairs. A substantial volume of data serves as input, and with the use of AI and preventative maintenance solutions, data points and meaningful conclusions are generated as output. The entire process assists in mitigating an issue prior to its emergence. Fuel efficiency is a critical statistic for aircraft OEMs, perhaps enhanced with artificial intelligence. The production of lightweight aircraft components enables even little enhancements in fuel efficiency to substantially impact the aircraft's emissions. AI is aiding pilots during flights by evaluating essential data such as the fuel system, system status, weather conditions, and other pertinent factors that may be analysed in real-time to enhance flight paths. AI facilitates enhanced human-machine collaboration in the aerospace sector by reducing time-intensive processes.

GENERATIVE DESIGN

In the aerospace industry, artificial intelligence is utilised to discover innovative methods for designing components that are more efficient, faster, and lighter. Generative AI is advancing markedly in the aerospace sector, particularly in the creation of images and films. Engineers and designers can employ it to produce precise and comprehensive 3D models of aeroplane components, hence facilitating the production process. This facilitates enhanced quality control, expedited prototyping, and potentially reduced production costs. Moreover, generative AI will facilitate virtual reality training simulations that will enhance the proficiency and safety of pilots and crew members. Machine learning techniques are being employed to produce innovative product designs that align with current demand. Engineers and product designers can more easily ascertain the ideal design due to the multitude of choices accessible within a brief timeframe.

EFFECTIVE SUPPLY CHAIN ADMINISTRATION

The aviation industry is optimising operations through the use of AI into the supply chain. Enhanced supply chain efficiency facilitates equipment maintenance and routine repairs more effectively than manual methods. This also conserves financial resources and minimises downtime, as the timing for maintenance is predetermined. Enhancing the efficacy of supply chain management through automated data collection is straightforward.

ENHANCED QUALITY ASSURANCE

The primary objective of quality assurance is to ensure that a product or service consistently adheres to the established quality standards. This is achieved by meticulously monitoring each phase of the production process. An autonomous AI solution can automate quality assurance, resulting in significant time and cost savings. Machine learning has enhanced the rate of fault detection by nearly 90% through the automation of quality testing. AI-enhanced simulators, integrated with virtual reality systems, can enhance pilot training facilities. Pilots will obtain a more realistic simulation experience. These simulators can also be utilised to collect and assess diverse training-related data to generate tailored training data that employs biometrics to track a user's success.

AUTOMATION IN THE COCKPIT

Artificial Intelligence and Machine Learning are transforming the domain of autopilot. Nonetheless, these technologies extend beyond data mining to offer predictive analytics, facilitate decision-making in intricate scenarios, or even assume control of certain phases of flight. This is accomplished through the integration of AI and ML algorithms with existing cockpit technologies, providing pilots with enhanced situational awareness, customised workflows, and improved decision-making capabilities. The automation of cockpits, resulting from the aerospace industry's embrace of these advancements, serves as a basis for the creation of safer, more efficient, more technologically sophisticated aviation systems.

ENHANCED SAFETY AND COMPLIANCE

To enhance aircraft development through automation, AI may generate models of human behaviour and actions. It can also examine aeroplane specifications. This can mitigate human mistake and enhance safety. It resembles an automatic evaluation. The aircraft sector is well-acquainted with autonomy. It is relatively common and straightforward for an aircraft to autonomously execute takeoff and landing manoeuvres. The challenge is in automating air traffic control systems, which now encounter numerous challenges. The intricacy of air traffic control systems can only be managed by AI and machine learning without presenting significant safety concerns. This is an essential initial step in encouraging individuals to choose a transition towards increased autonomy.

AEROSPACE DEVELOPMENT, SIMULATION, AND EVALUATION

System modelling in the aerospace sector extensively incorporates artificial intelligence. However, not all simulations necessitate direct human engagement to operate. Besides assisting in the development of hardware models, software, and simulation environments, AI may also forecast particular types of codes. This combination enhances productivity in the development and testing process while also decreasing delivery time. Moreover, it enhances precision as machines often excel at detecting faults compared to human engineers. Physical prototyping is not a viable option for testing new aircraft systems due to substantial margins. An accident in a densely populated area presents a significant danger and incurs expenditures amounting to hundreds of millions of dollars for each fatality. Engineers can transfer testing to a virtual environment utilising AI and machine learning, hence enhancing the probability of a model's success prior to the creation of a physical prototype.

PROGRESS IN BIOMETRICS AND SECURITY

Digital identity and biometrics have become a standard component of our annual trends report. The implementation of biometrics facilitates the enhancement of customer experience and the augmentation of security in airports and, ultimately, across borders. The worldwide implementation of digital identity and biometric technology is rapidly increasing. Travellers can efficiently transit each stage of the procedure, including self-boarding gates, pre-security automated gates, and enrolment at kiosks or counters, by solely scanning their faces using biometric technology. Airports and airlines are increasingly using biometric technology to ensure a seamless travel experience from kerb to gate, with widespread implementation anticipated by 2025.

TECHNOLOGIES FOR SUSTAINABLE AVIATION

These solutions are expected to significantly reduce greenhouse gas emissions while maintaining operational efficiency. Hydrogen fuel cells are being seen as a feasible and transformative energy source. Utilising hydrogen as a fuel enables aircraft to operate with zero pollution. Despite existing challenges, including as the establishment of infrastructure for hydrogen production, storage, and delivery, collaborative efforts throughout the industry are accelerating progress. In addition to propulsion, hydrogen may energise auxiliary systems and airport operations. For the first time, we are proactively reevaluating aviation's role in a cleaner, more sustainable world rather than simply mitigating its environmental impact. Sustainable Aviation Fuel (SAF), a sustainable alternative to traditional jet fuel, is a crucial innovation in the travel industry. Adoption is essential for attaining global climate objectives and meeting the increasing demand for sustainable transport alternatives. To guarantee enduring viability and expansion, SAF is transforming the industry's sustainability initiatives by advocating for technology advancements in fuel production and a circular economy.

IN-FLIGHT PERSONALISATION CONNECTIVITY AND ENTERTAINMENT

In-flight entertainment and connection are increasingly essential components of the traveler's experience. Airlines are investing in advanced satellite technology to provide high-speed internet connectivity on flights, as passengers increasingly want to remain connected while airborne. This enhances the whole flight experience by allowing passengers to work, stream content, and sustain connections with loved ones. Comprehensive customisation at each phase of the journey is increasingly anticipated by travellers. Customers want their travel experience to evolve seamlessly, contextually, and intuitively with technological advancements. Moreover, airlines are leveraging data analytics to tailor in-flight entertainment options to individual passenger tastes. This change aims to provide airlines with valuable insights into passenger behaviour.

INTERNET OF THINGS AND PRIVATE WIRELESS NETWORKS

The technological transformation occurring in airports and airlines is focused on the Internet of Things (IoT). The basic structures must undergo substantial alteration to manage the immense volume of data produced by robotics and the Internet of Things. This encompasses the efficient use of high-speed, low-latency networks for real-time data transmission, edge computing frameworks for processing data near its source, and robust cloud infrastructure for data storage and analysis. Moreover, enhancing cybersecurity is crucial for safeguarding this interconnected environment from potential threats. Airports and airlines are investing in scalable and adaptive IT systems capable of accommodating the increasing complexity and requirements of IoT applications, alongside the technological infrastructure. These infrastructure enhancements ensure that IoT technologies function seamlessly and foster innovation in the aviation sector. The benefits of Private Wireless Networks (PWNs) are increasingly evident, encompassing enhanced operational efficiency through improved coverage and superior network security. PWNs are crucial for airports and their collaborators exploring technologies such as driverless vehicles and the Internet of Things.

AUGMENTED REALITY, VIRTUAL REALITY, AND EXTENDED REALITY

The education of aviation professionals is undergoing a profound transformation due to immersive technologies like Augmented Reality (AR), Virtual Reality (VR), and Extended Reality (XR). Contemporary simulation technologies that precisely replicate real-world scenarios are now accessible to pilots, flight attendants, ground operations personnel, and technicians. Employees may practice emergency protocols, manage complex procedures, and respond to challenging events in a safe environment using VR-based training programmes. Similarly, technicians can utilise augmented reality overlays to obtain detailed schematics and sequential instructions during maintenance, therefore reducing errors and enhancing productivity. Besides reducing training costs, these immersive technologies enhance confidence and skill retention, thereby better preparing aviation personnel to confront real-world challenges.

CONCLUSION

In conclusion, the integration of AI and ML with aerospace and military has inaugurated a new era of possibilities. These technologies, encompassing autonomous systems and data-driven business models, have fundamentally transformed the corporate landscape. Nonetheless, ethical dilemmas and security concerns underscore the imperative of responsible innovation. The integration of human cognition with AI and machine learning anticipates substantial advancements in the future. New frontiers, cooperative frameworks, and quantum computing all invite exploration.

REFERENCES

Barkalov, K., Lebedev, I., Usova, M., Romanova, D., Ryazanov, D., & Strijhak, S. (2022). Optimization of turbulence model parameters using the global search method combined with machine learning. *Mathematics*, *10*(15), 2708. DOI: 10.3390/math10152708

Beetham, S., & Capecelatro, J. (2020). Formulating turbulence closures using sparse regression with embedded form invariance. *Physical Review Fluids*, *5*(8), 084611. DOI: 10.1103/PhysRevFluids.5.084611

Bounds, C. P., Uddin, M., & Desai, S. (2023). Tuning of Turbulence Model Closure Coefficients Using an Explainability Based Machine Learning Algorithm (No. 2023-01-0562). SAE Technical Paper.

Brenner, M. P., Eldredge, J. D., & Freund, J. B. (2019). Perspective on machine learning for advancing fluid mechanics. *Physical Review Fluids*, *4*(10), 100501. DOI: 10.1103/PhysRevFluids.4.100501

Brunton, S. L., & Kutz, J. N. (2022). *Data-driven science and engineering: Machine learning, dynamical systems, and control*. Cambridge University Press. DOI: 10.1017/9781009089517

Dangeti, C. S. (2018). Sensitivity of Turbulence Closure Coefficients on the Aerodynamic Predictions of Flow over a Simplified Road Vehicle (Master's thesis, The University of North Carolina at Charlotte).

Eivazi, H., Le Clainche, S., Hoyas, S., & Vinuesa, R. (2022). Towards extraction of orthogonal and parsimonious non-linear modes from turbulent flows. *Expert Systems with Applications*, *202*, 117038. DOI: 10.1016/j.eswa.2022.117038

Godunov, S. K., & Bohachevsky, I. (1959). Finite difference method for numerical computation of discontinuous solutions of the equations of fluid dynamics. Matematičeskij sbornik, 47(3), 271-306.

Guilmineau, E., Deng, G. B., Leroyer, A., Queutey, P., Visonneau, M., & Wackers, J. (2018). Assessment of hybrid RANS-LES formulations for flow simulation around the Ahmed body. *Computers & Fluids*, *176*, 302–319. DOI: 10.1016/j.compfluid.2017.01.005

Jeon, J., Lee, J., & Kim, S. J. (2022). Finite volume method network for the acceleration of unsteady computational fluid dynamics: Non-reacting and reacting flows. *International Journal of Energy Research, 46*(8), 10770–10795. DOI: 10.1002/er.7879

Kraichnan, R. H. (1967). Inertial ranges in two-dimensional turbulence. *The Physics of Fluids, 10*(7), 1417–1423. DOI: 10.1063/1.1762301

Ling, J., Kurzawski, A., & Templeton, J. (2016). Reynolds averaged turbulence modelling using deep neural networks with embedded invariance. *Journal of Fluid Mechanics, 807*, 155–166. DOI: 10.1017/jfm.2016.615

Liu, W., Fang, J., Rolfo, S., Moulinec, C., & Emerson, D. R. (2021). An iterative machine-learning framework for RANS turbulence modelling. *International Journal of Heat and Fluid Flow, 90*, 108822. DOI: 10.1016/j.ijheatfluidflow.2021.108822

Maulik, R., San, O., Rasheed, A., & Vedula, P. (2019). Subgrid modelling for two-dimensional turbulence using neural networks. *Journal of Fluid Mechanics, 858*, 122–144. DOI: 10.1017/jfm.2018.770

Morita, Y., Rezaeiravesh, S., Tabatabaei, N., Vinuesa, R., Fukagata, K., & Schlatter, P. (2022). Applying Bayesian optimization with Gaussian process regression to computational fluid dynamics problems. *Journal of Computational Physics, 449*, 110788. DOI: 10.1016/j.jcp.2021.110788

Obiols-Sales, O., Vishnu, A., Malaya, N., & Chandramowliswharan, A. (2020, June). CFDNet: A deep learning-based accelerator for fluid simulations. In *Proceedings of the 34th ACM international conference on supercomputing* (pp. 1-12). DOI: 10.1145/3392717.3392772

Pope, S. B. (1999). A perspective on turbulence modelling. In *Modeling complex turbulent flows* (pp. 53–67). Springer Netherlands. DOI: 10.1007/978-94-011-4724-8_5

Poroseva, S. V., Colmenares, F. J. D., & Murman, S. M. (2016). On the accuracy of RANS simulations with DNS data. *Physics of Fluids, 28*(11), 115102. DOI: 10.1063/1.4966639 PMID: 30271109

Shan, T., Tang, W., Dang, X., Li, M., Yang, F., Xu, S., & Wu, J. (2020). Study on a fast solver for Poisson's equation based on deep learning technique. *IEEE Transactions on Antennas and Propagation*, *68*(9), 6725–6733. DOI: 10.1109/TAP.2020.2985172

Shi, H. Y., Wang, S., & Wang, P. Y. (2024). From simulation to reality: CFD-ML-driven structural optimization and experimental analysis of thermal plasma reactors. *Journal of Environmental Chemical Engineering*, *12*(3), 112998. DOI: 10.1016/j.jece.2024.112998

Spalart, P., & Allmaras, S. (1992, January). A one-equation turbulence model for aerodynamic flows. In 30th aerospace sciences meeting and exhibit (p. 439). DOI: 10.2514/6.1992-439

Spalart, P. R. (2000). Strategies for turbulence modelling and simulations. *International Journal of Heat and Fluid Flow*, *21*(3), 252–263. DOI: 10.1016/S0142-727X(00)00007-2

Vinuesa, R., Hosseini, S. M., Hanifi, A., Henningson, D. S., & Schlatter, P. (2017). Pressure-gradient turbulent boundary layers developing around a wing section. *Flow, Turbulence and Combustion*, *99*(3-4), 613–641. DOI: 10.1007/s10494-017-9840-z PMID: 30069159

Wang, B., & Wang, J. (2021). Application of artificial intelligence in computational fluid dynamics. *Industrial & Engineering Chemistry Research*, *60*(7), 2772–2790. DOI: 10.1021/acs.iecr.0c05045

Weller, H. G., Tabor, G., Jasak, H., & Fureby, C. (1998). A tensorial approach to computational continuum mechanics using object-oriented techniques. *Computers in Physics*, *12*(6), 620–631. DOI: 10.1063/1.168744

Wu, J., Xiao, H., Sun, R., & Wang, Q. (2019). Reynolds-averaged Navier–Stokes equations with explicit data-driven Reynolds stress closure can be ill-conditioned. *Journal of Fluid Mechanics*, *869*, 553–586. DOI: 10.1017/jfm.2019.205

Yarlanki, S., Rajendran, B., & Hamann, H. (2012, May). Estimation of turbulence closure coefficients for data centers using machine learning algorithms. In 13th InterSociety Conference on Thermal and Thermomechanical Phenomena in Electronic Systems (pp. 38-42). IEEE. DOI: 10.1109/ITHERM.2012.6231411

Zhu, L., Zhang, W., Kou, J., & Liu, Y. (2019). Machine learning methods for turbulence modelling in subsonic flows around airfoils. *Physics of Fluids*, *31*(1), 015105. DOI: 10.1063/1.5061693

Chapter 3
Transforming Aerospace Design and Manufacturing Through Machine Learning

Kishorebabu Dasari
https://orcid.org/0000-0001-6920-0674
Keshav Memorial Institute of Technology, India

Sujana Parry
MTAR Technologies Limited, India

Srinivas Mekala
KG Reddy College of Engineering and Technology, India

ABSTRACT

Machine learning (ML) is driving change in aerospace design and manufacturing in order to increase efficiency, safety, and innovation. Complex designs and requirements required in aerospace industry. ML can improve design quality, support design thinking by generating multiple options, and enhance predictive analytics to detect potential problems. ML makes manufacturing processes smarter by predicting demand and managing resources effectively, increasing operational efficiency and improving product quality. ML helps to prevent errors and increase safety and secu-

DOI: 10.4018/979-8-3693-7525-9.ch003

rity through predictive maintenance, non-destructive testing, and anomaly detection. ML can handle challenges such as data collection, interpreting in a safety-critical business context. ML offers great opportunities, from AI-powered innovation and human-machine collaboration to the potential impact of quantum computing in solving complex aerospace problems. Continued advances in machine learning will help to shape the future of aerospace, making machines more efficient, safer, and more innovative.

INTRODUCTION

The aerospace industry (Caliari *et al.,* 2023) is one of the most complex, demanding and high-risk industries in the global economy. Aircraft and its components must meet the latest safety, security and operational standards. Every phase of aerospace manufacturing, from initial design to final production and assembly, requires precision. The design of aerospace systems must take into account not only the mechanical and process complexity, it also the harsh environmental conditions to which these systems will be exposed, such as weather, temperature, pressure and high-speed dynamics. Aerospace manufacturing involves the use of a variety of specialty materials, including composites and alloys, to ensure that the product can withstand operational stresses while minimizing weight and increased workload.

Regulatory bodies (Hartl *et al.,* 2015) such as the Federal Aviation Administration (FAA) in the United States and the European Union Aviation Safety Agency (EASA) in Europe have strict rules that govern all aspects of aviation life, from design and testing to production and maintenance. Compliance with these strict regulations is essential to ensuring the safety and reliability of the aircraft and passenger safety and mission success.

Innovation is the driving force of the aviation industry, with companies and researchers eager to create more efficient, lighter, and more environment friendly aircraft. Balancing innovation with strict regulatory compliance poses a unique challenge for aviation engineers and manufacturers. Achieving both goals requires not only top-notch expertise, but also new technologies that can increase accuracy, reduce costs, and speed up rapid growth.

Machine learning (ML) (Brunton *et al.*, 2021) is considered a powerful tool to solve these problems. ML has the ability to analyze large amounts of data, identify patterns, and make accurate predictions. These resources are especially valuable in industries like aviation, where even a small design or manufacturing error can cause catastrophic damage. ML ability to improve processes, develop design models, and predict outcomes is reshaping the way aircraft design and manufacturing are conducted.

In the design world, ML has enabled aerospace engineers to improve design quality faster and more accurately than ever before. Design models (Bertoni *et al.*, 2020) often require simulation and prototyping, which is time-consuming and costly. ML algorithms can automate many of these processes by analyzing previously generated data and predicting which configurations will perform best. This optimization is especially important in areas such as aerodynamics, integrity, and material selection, where performance and efficiency can be improved with small changes. Example, ML can help determine the best materials for building a lightweight yet strong aircraft wing, improving fuel efficiency and overall performance. Similarly, in the design of an aircraft, ML models can predict how various materials and designs will perform in airspace, allowing engineers to choose the best options without much physical testing.

In the manufacturing industry (Zhan & Li, 2021), ML is driving the advancement of smarter and more efficient production systems. 3D printing, also known as additive manufacturing, has revolutionized the aerospace industry by enabling the production of intricate components that would be otherwise unattainable through conventional manufacturing methods. ML enhances this process by continuously monitoring and adjusting parameters in real-time to maximize material usage, minimize waste, and enhance the precision of each printed part. ML is transforming quality control in the aerospace industry. Traditional quality control methods heavily depend on manual inspections, which are not only time-consuming but also susceptible to human mistakes. Systems based in Malaysia, especially those utilizing computer vision and image recognition, have the capability to examine components at a microscopic level, identifying defects that may not be visible to the naked eye. This automated quality control guarantees that each component adheres to the rigorous safety and performance standards necessary for aerospace applications.

Enhancing design and manufacturing processes, ML is playing a crucial role in guaranteeing operational safety. One of the most promising applications of ML in aerospace is the ability to predict maintenance needs. Today's airplanes are equipped with numerous sensors that gather real-time data on various aspects, such as engine performance and airframe stress. ML algorithms can examine this data to detect patterns that indicate potential failures or maintenance requirements before they escalate into critical situations. By anticipating when and where maintenance is necessary, airlines and aerospace manufacturers can greatly minimize downtime, decrease maintenance expenses, and enhance safety. This is especially important in the aerospace industry, where unexpected maintenance can result in flight delays, mission failures, or even accidents.

The incorporation of ML into aerospace is not without its obstacles. One of the primary challenges (Lee, 2023) is the accessibility and reliability of data. Aerospace systems produce massive quantities of data, but a significant portion of it remains unstructured or isolated within various departments and organizations. Efficiently managing, analyzing, and utilizing this data necessitates advanced data management systems and a collaborative approach to dismantle these silos. Furthermore, the “black box” characteristic of many ML algorithms raises concerns about transparency and reliability. In industries where safety is paramount, such as aerospace, decision-makers need to comprehend and provide valid justifications for the suggestions generated by machine learning systems. This is particularly crucial in regulatory settings, where every decision must be properly documented and validated to guarantee adherence to safety regulations.

Regardless of these challenges, the ability use of ML in aerospace is vast. The era offers a manner to make design and production processes extra green, lessen mistakes, and enhance safety. As ML maintains to conform, it's miles likely to play a fair more principal position in the aerospace industry, driving innovations that had been previously unimaginable. The combination of ML with rising technology like quantum computing should free up new opportunities for solving complicated design troubles which can be currently beyond the skills of even the most advanced computer systems. ML ought to enable the improvement of autonomous plane and spacecraft that could perform without human intervention, increasing the frontiers of aerospace innovation.

Machine learning is positioned to revolutionize the aerospace sector. By automating data analysis, optimizing intricate designs, increasing manufacturing precision, and enhancing operational safety, ML empowers the aerospace industry to tackle its most pressing challenges while driving innovation and pushing boundaries. This technology matures and gains broader acceptance, its influence on aerospace design and manufacturing will continue to expand, shaping the industry's future.

ML a specialized division of artificial intelligence (AI), is increasingly important in modern aerospace engineering. The aerospace industry involves vast, complex datasets created during different phases like design, testing, and production. Manually handling this significant data burden is not just time-consuming but also susceptible to human error. ML provides a solution by automating data analysis, allowing for real-time insights, and aiding in well-informed decision-making. These features have revolutionized how aerospace firms create airplanes, improve manufacturing processes, and guarantee prolonged operational dependability.

The aerospace field produces a lot of data (Arista *et al.,* 2022) from various sources like sensors, tests, and manufacturing processes. Old methods of analyzing this data have a hard time providing useful insights fast enough for real-time needs. Machine learning models can, however, handle these big datasets well, finding patterns and connections that people can't easily see. For example, machine learning can spot early indicators of equipment issues or design problems before they turn into expensive or risky situations.

Aerospace businesses can gain an edge by processing big data, as this helps them adapt to changing challenges and market needs. Machine learning models get better and better as they handle more data, which leads to more precise predictions and improved systems over time.

Key advantage of ML is that it automates complex analysis, which speeds up and improves decision-making. During the design stage, engineers can utilize ML algorithms to conduct several simulations to determine optimal configurations for things like aerodynamic efficiency, fuel use, and structural strength. These simulations, typically done in real-time, assist engineers in making informed decisions with current data and help prevent expensive design mistakes.

Combination with real-time sensor inputs, production processes can be overseen by ML algorithms, which is a great advancement in the manufacturing field. Example, ML related automated systems with an image recognition approach are able to check every component for defects. This instant feedback enables the manufacturer to resolve the problem at once, reducing delays and wasting a minimum amount of resources. ML techniques optimize manufacturing and design activity successive enhanced speed and accuracy in decision-making.

Disparate aerospace systems that include engines, avionics and control surfaces are very intricate and call for thorough examinations. The well-trained ML models do well in pattern recognition so that engineers can discern subtle irregularities and patterns hidden in the test data that can be considered as prospective faults. These insights assist in component analysis increasing reliability of some of the components against unexpected failures while in use.

ML enhanced predictive modeling has a critical impact on both the design and the servicing of aerospace systems. By using predictive models, there is an opportunity to anticipate how a system will operate within a particular environment and make necessary design alterations before the actual prototype is constructed. In the operational stage of the product, there are predictive maintenance models that use live sensor data to predict possible failure conditions and times for a component or system. This approach helps aerospace firms minimize unscheduled maintenance, increase the equipment's life cycle, and decrease operating expenses.

Through the use of MM across the entire aerospace lifecycle (Goksidan, 2019), great savings can be made. First operational tasks are being automated, reducing the amount of human effort required and decreasing the level of possible human mistakes. Second, the possibility of running thousands of virtual simulations in a much shorter timeframe than the physical testing allows for the lowering of development costs. Furthermore, ML-enhanced predictive maintenance reduces equipment downtime, as well as expenses related to breakdowns and repairs.

ML makes it possible to continuously optimize performance. Lighter, stronger, and more efficient aircraft are the result of design iterations guided by insights from machine learning. ML-powered intelligent automation on production lines reduces waste and maintains high quality standards. As a result, aerospace firms can use less resources to provide greater results, promoting environmental and economic sustainability.

As the aerospace sector develops further, machine learning's impact is anticipated to increase. Real-time decision-making and high levels of reliability will become even more important as autonomous flying systems and space exploration become more common. Future advancements in quantum computing could expand ML's potential and enable it to address even more difficult design and production problems.

ML will be essential to the optimization (Xue et al., 2023) of these new systems as aerospace businesses move toward greener technologies like electric propulsion and hydrogen-powered planes. Simultaneously, collaborative design environments—where engineers and machine learning algorithms collaborate—will proliferate, resulting in even more inventive solutions.

In order to handle the complexity of aerospace design and manufacturing, machine learning has emerged as a vital technology. Aerospace businesses may increase performance, lower costs, and improve reliability thanks to its capacity to process enormous volumes of data, automate decision-making, identify patterns, and offer predictive insights. The use of machine learning (ML) technology will be crucial as the aerospace sector develops to fulfill the demands of efficiency, accuracy, and time. Building smarter, safer, and more sustainable aircraft and systems by utilizing machine learning's full potential is the key to the future of aerospace engineering.

MACHINE LEARNING IN AEROSPACE DESIGN

The aircraft sector requires designs (Ni, 2022) to be efficient, safe, and precise. Aircraft design has always depended on iterative testing and improvement, which entails building physical prototypes and simulations to verify design ideas. These techniques work well, but they are expensive and time-consuming. This procedure has been completely transformed by ML, which improves design workflows and automates intricate analysis. Engineers may more effectively build better-performing airplanes by using ML-driven techniques for everything from aerodynamic optimization to predictive analytics.

Optimization of Aircraft Design

Manual procedures involving numerous physical testing iterations and comprehensive simulations to contemporary workflows powered by machine learning algorithms, aircraft design optimization has seen significant change. Optimizing performance indicators including speed, fuel efficiency, structural strength, and environmental sustainability is the goal.

Aerodynamic Optimization (Li, Du, & Martins, 2022): To determine the most aerodynamically efficient designs, machine learning (ML) algorithms—in particular, reinforcement learning and genetic algorithms—analyze massive datasets from computational fluid dynamics (CFD) simulations. In order to direct engineers toward more efficient forms for wings, fuselages, and control surfaces, these algorithms rapidly learn which design elements decrease drag, reduce turbulence, and improve lift.

Material Selection and Configuration (Leonard, Prasetiyo, & Putra, 2023): Selecting acceptable materials for airplane components entail balancing trade-offs between weight, durability, and cost. ML-based optimization techniques examine material attributes and usage patterns to recommend optimal designs. Machine learning, for instance, can identify applications for lightweight composites without sacrificing structural integrity.

Wing optimization (Elham, 2015) is a prominent use of machine learning in aircraft design. Fuel efficiency, stability, and payload capacity are all directly impacted by wing shape and configuration. By lowering drag, ML models trained on CFD and historical flight data can suggest wing designs that increase fuel efficiency. By doing away with the requirement for several physical prototypes, this approach speeds up development cycles and saves a substantial amount of money.

Generative Design

A state-of-the-art method called generative design (Warey *et al.,* 2023) uses machine learning algorithms to automatically produce a number of design options based on predetermined parameters including strength, weight, material availability, and environmental factors. Through this technique, the design paradigm is changed from manually producing a small number of models to investigating a large number of potential solutions.

Engineers enter design specifications such as maximum weight, necessary load-bearing capacity, or aerodynamic performance into an ML algorithm in a generative design system. Hundreds or thousands of possible design solutions that satisfy these requirements are subsequently produced by the algorithm. The most intriguing solutions can be reviewed and refined by engineers.

Generative design (Sun *et al.,* 2024) has a lot to offer the aerospace sector since it makes it possible to create intricate, lightweight structures that were previously difficult to build by hand. Because every kilogram saved translates to increased fuel economy and lower emissions, it is crucial to reduce weight without compromising strength, particularly in the development of airplanes and spacecraft. With generative design, the best possible design is produced with the least amount of material waste.

Spacecraft parts that need to withstand high stress while weighing as little as possible, including brackets and support structures, have been developed using generative design. For instance, NASA developed lightweight aluminum brackets for spacecraft using generative design driven by machine learning. Compared to its traditionally built equivalents, these components are 50% lighter while maintaining structural strength.

Predictive Analytics in Design

The reliability of aircraft systems is greatly increased by predictive analytics, which uses machine learning models to foresee possible problems during the design stage. Predictive models (Dube & Hiravennavar, 2020) find design flaws prior to physical testing by modeling long-term use scenarios and examining historical data. This proactive strategy lowers the chance of recalls, increases safety, and helps prevent expensive mistakes.

Over time, fatigue and failure may result from the frequent stress that takeoffs, landings, and environmental factors place on aerospace components. By modeling these stresses under varied operating settings, machine learning-based predictive analytics may predict how components will deteriorate over time. This prolongs component life and prevents in-service failures by enabling engineers to make the required changes during the design phase.

Performance deterioration is also predicted using predictive analytics. For instance, ML models that have been trained on sensor data from prior flights are able to forecast when avionics systems or engines would start to function poorly. By using these projections, engineers can create systems that are more robust against deterioration, increasing overall dependability.

Case Study: Predictive Modeling to Avoid Recalls

In the past, when unexpected design problems were found after deployment, aircraft makers had to deal with costly recalls and retrofitting campaigns. By anticipating failure areas, predictive analytics lessens (Nogueira *et al.*, 2023) the possibility of such situations. For instance, engineers can choose different materials or modify the design before mass production starts if a predictive model indicates that a certain part is likely to wear under high humidity circumstances.

In aircraft design, where traditional workflows are being transformed by optimization, generative design, and predictive analytics, machine learning has become essential. With the use of these technologies, engineers can investigate novel designs, use less material, forecast long-term performance, and guarantee the efficiency and safety of aircraft. By accelerating the design process and minimizing errors, ML helps aerospace companies maintain a competitive edge in an industry where performance, safety, and sustainability are paramount. As ML technologies continue to evolve, its impact on design will only deepen, paving the way for the next generation of aerospace innovations.

MACHINE LEARNING IN AEROSPACE MANUFACTURING

Aerospace manufacturing (Jiang, Tran, & Williams, 2023) is changing quickly, and machine learning (ML) is essential to streamlining operations, increasing quality, and optimizing production processes. ML technologies help businesses reach high levels of precision, efficiency, and scalability in a variety of processes, from supply chain optimization to the exact creation of components via additive manufacturing. This chapter examines the ways in which machine learning (ML) improves supply chain management, robotics-driven automation, automated quality control, and smart manufacturing, among other areas of aerospace production.

Smart Manufacturing with ML

3D printing, often known as additive manufacturing, is revolutionizing the aerospace sector. It enables producers to make intricate, lightweight structures that were previously impossible or difficult to accomplish with conventional manufacturing techniques. By increasing accuracy and lowering errors in real-time, ML integration into these procedures elevates the advantages of AM to new levels.

Real-Time Process Optimization: While 3D printing, ML algorithms keep an eye on production parameters (Ma *et al.,* 2023) including temperature, pressure, and material flow. To ensure constant quality, the ML system conducts real-time modifications if the sensors built into the apparatus identify departures from ideal values. To guarantee homogeneity in every layer of a component, for example, it can optimize cooling periods and layer deposition.

Error Detection and Correction: Time and materials may be wasted in traditional manufacturing when errors are not discovered until the end of the production cycle. In order to anticipate such flaws early in the process, machine learning-driven systems continuously examine sensor data. The technology reduces waste and guarantees constant quality by proactively making modifications, such as changing printing speeds or recalibrating the machinery.

Improved Build Quality and Material Efficiency: Machine learning models are able to forecast the ideal quantity of material needed for every print, which minimizes waste and guarantees that the finished product satisfies specifications for strength and durability. This is particularly crucial in the aerospace industry, where it's essential to reduce weight without sacrificing structural integrity.

The fabrication of aviation engine components is an instance of additive manufacturing aided by machine learning. ML makes sure that these intricate, high-stress parts are printed with few flaws through smart manufacturing, which eliminates the need for rework or post-production inspections. Higher-quality outputs, reduced costs, and quicker production cycles are the outcomes of this.

Automated Quality Control

Quality control (Javaid *et al.*, 2021) is an essential component of the manufacturing process in the aerospace industry because even the smallest flaw can result in catastrophic failures. Automated quality control systems driven by machine learning (ML) use computer vision and image recognition to examine components with a high degree of precision, much outperforming manual inspections.

ML-Based Image Recognition Systems: These systems train machine learning models that can accurately detect flaws using enormous datasets of photos from prior errors. High-resolution cameras take pictures at different points during the manufacturing process of components. To find anomalies like cracks, inconsistent material, or deformities, the machine learning algorithm compares these photos to reference datasets.

Finding Microscopic Flaws: Although some flaws, such as microcracks, are too tiny to be seen with the naked eye, they can jeopardize the dependability and safety of aeronautical parts. Only parts that satisfy the highest quality standards move on to final assembly thanks to ML systems' accurate analysis of these minute flaws.

For instance, the production of composite materials for aero plane fuselages uses automated quality control systems. As the composites are laid, these systems examine the layers to look for irregularities that might compromise the structure. This increases the final product's durability and safety by guaranteeing that every layer is perfect.

Supply Chain Optimization

With several supplier layers, strict deadlines, and geographically scattered activities, the aircraft manufacturing supply chain (Falcão e Cunha *et al.,* 2023) is extremely complex. Through demand forecasting, inventory optimization, and predictive maintenance, ML provides substantial benefits in handling this complexity.

Inventory optimization: To forecast future component demand, machine learning algorithms examine market patterns and historical data. By doing this, producers are able to keep its inventory levels at ideal levels and prevent shortages and overstocks. Accurate inventory management (Yang *et al.,* 2018) is crucial to lowering costs and guaranteeing on-time production in the aerospace industry, because many parts are costly and have lengthy lead times.

Demand Forecasting: Seasonal patterns, geopolitical developments, and new technology all influence the demand that aerospace companies must deal with. A variety of data inputs can be processed by ML models to forecast changes (Dodin *et al.,* 2023) in demand, enabling producers to proactively modify its production plans. By ensuring that parts are available when needed, production delays are minimized and delivery deadlines are met.

Predictive Maintenance of Equipment: By anticipating when manufacturing equipment may need maintenance, ML significantly reduces unplanned downtime. Machine learning algorithms (Zhong *et al.,* 2023) examine continually collected sensor data from machines to find trends that might point to equipment failure. Manufacturers can prevent expensive interruptions and sustain high levels of output by proactively scheduling maintenance.

Predictive maintenance, for instance, is used by aerospace firms to keep an eye on the functioning of robotic welders and CNC machines. Businesses avoid unscheduled downtime and guarantee a seamless supply chain operation by anticipating when these machines are likely to require maintenance.

Robotics and Automation

By increasing productivity, accuracy, and scalability, robotics and automation driven by machine learning have revolutionized the aerospace industry. Compared to conventional human-operated systems, robots with sophisticated machine learning algorithms can complete repetitive activities like welding, riveting, and assembly more accurately and quickly.

ML-Driven Robotics in Assembly Lines: By continuously learning (Acuña, 2021) from previous operations, robots can increase consistency and decrease production times. For instance, riveting robots can modify its methods to fit different kinds of joints and materials, guaranteeing excellent performance on a range of parts.

Collaborative Robots: Collaborative robots are used by certain aerospace manufacturers to assist human operators. These robots can adapt its behavior by learning from human behavior thanks to machine learning algorithms. While retaining the flexibility to carry out intricate operations that call for human skill, increase productivity.

Autonomous Drilling and Welding: Accurate drilling and welding are crucial for maintaining component integrity in aircraft manufacturing. Robots with machine learning capabilities can complete these jobs on its own with little guidance, guaranteeing excellent accuracy. And they are able to identify possible problems during drilling or welding and make adjustments in real time to prevent flaws.

Adaptation and Continuous Learning: The capacity of ML-powered robotics to learn from past tasks is one of its main advantages. Robots can grow more productive over time thanks to this ongoing learning, which lowers manufacturing costs and raises the standard of output overall.

The process of assembling aeroplane wings is an illustration of automation powered by robotics. ML-aligned robots perform jobs like drilling thousands of fastener holes with remarkable accuracy. To ensure that every hole is drilled precisely, these robots can modify its activities according to the wing's material qualities. This degree of accuracy guarantees that the finished product satisfies strict aerospace standards and minimizes rework.

ENHANCING SAFETY AND QUALITY ASSURANCE

In the aerospace sector, where even little flaws or system failures can have disastrous results, safety and quality control are crucial. By facilitating anomaly identification, non-destructive testing, and predictive maintenance, ML technologies are essential to enhancing these fields. By proactively managing equipment health, detecting internal faults, and identifying abnormalities in real-time, aerospace businesses can ensure optimal safety and operational reliability.

Predictive Maintenance

Predictive maintenance is a proactive strategy that uses ML to analyze real-time sensor data from aircraft engines, airframes, and critical systems in order to predict equipment problems. Predictive maintenance guarantees that parts are repaired only when required, reducing unplanned breakdowns and prolonging equipment life cycles, in contrast to traditional maintenance methods that depend on planned inspections or reactive repairs.

Real-Time Monitoring of Equipment Health: Aircraft generate huge volumes of data through sensors implanted in engines, hydraulic systems, and other components. ML models continuously examine this data to find patterns indicating wear, corrosion, or other symptoms of degradation. For instance, an engine's vibration data may show the early signs of bearing wear that may eventually cause the engine to fail.

Optimized Maintenance Scheduling: By predicting when specific parts are likely to fail, ML-based predictive maintenance helps airlines and manufacturers optimize maintenance schedules. This lowers the chance of in-flight malfunctions, guarantees that spares are available when needed, and minimizes aircraft downtime.

Extended Equipment Life Cycles: Predictive maintenance not only reduces unexpected failures but also extends the life of critical components. Businesses may prevent needless part replacements and guarantee optimal equipment utilization by scheduling maintenance at the ideal time—before significant damage has occurred.

For example, aeroplane engines are frequently monitored using ML-based predictive maintenance, where early anomaly detection can avert catastrophic breakdowns and guarantee safe flight operations. For modern airlines looking to improve safety, cut expenses, and preserve operational effectiveness, this proactive strategy has become crucial.

Non-Destructive Testing (NDT)

A crucial procedure for examining the interior integrity of components without causing harm is non-destructive testing, or NDT (Torbali, Zolotas, & Avdelidis, 2023). It guarantees that vital components fulfil performance and safety standards. By more efficiently evaluating sensor data and identifying structural abnormalities that conventional techniques might overlook, ML models have improved NDT.

Better Internal Defect Detection: Machine learning algorithms are able to examine data from nondestructive testing (NDT) techniques including radiography, ultrasonic testing, and infrared thermography in order to detect internal defects like fatigue, corrosion, and cracks. ML models can identify tiny anomalies based on previous data and trends, even if these flaws might not be apparent during eye examinations.

Faster and More Accurate Inspections: Conventional NDT techniques can take a long time and need to be interpreted by a specialist. Systems with machine learning capabilities automate the study of inspection data, increasing accuracy and cutting down on inspection times. This guarantees that even the smallest flaws are found and fixed before they jeopardize the component's functionality or safety.

Removing the Need for Disassembly: ML-enhanced NDT saves time and money by enabling comprehensive inspections without the need to disassemble complicated systems. Destructive testing techniques are no longer necessary since sensors positioned on the surface of aeroplane wings or fuselages, for instance, may collect data that machine learning algorithms can use to evaluate structural integrity.

Turbine blade inspection is one instance of ML-enhanced NDT in aerospace. Early diagnosis of cracks and fatigue is crucial since these components are exposed to high levels of stress and harsh temperatures. To ensure safe and effective engine running, machine learning models examine ultrasonic data to identify wear indicators and forecast when the blades will require maintenance or replacement.

Anomaly Detection

ML-powered anomaly detection techniques are essential for spotting anomalies in operations, aircraft systems, and industrial processes. These algorithms pick up on the typical operation of machinery or systems and identify any deviations that might point to flaws or impending problems. Early detection of anomalies allows for swift corrective actions, preventing damage, accidents, or system downtime.

Manufacturing Process Anomaly Detection: Machine learning algorithms keep an eye on manufacturing machinery to guarantee constant production quality. The system notifies operators to look at and fix any anomalies (Dasari, Mekala, & Kaka, 2024) it finds, such as erratic welding patterns or uneven material thickness. This guarantees high-quality outputs and reduces the manufacture of defective parts.

Operational System Fault Detection: ML-based anomaly detection systems examine real-time sensor data from vital systems, like avionics and engines, during flight operations. The flight crew or ground engineers are alerted whenever an ML algorithm detects abnormal behaviour, such as sudden temperature spikes or pressure dips. This makes it possible to take preventative measures to stop more problems, such rerouting the aircraft or changing system settings.

Preventing Failures and Accidents: Preventing failures that could jeopardize flight safety requires early detection of anomalies. Anomaly detection systems, for instance, have been used to track fuel pump performance and spot warning indications of impending failure before the system breaks down. This proactive strategy guarantees safe operations and lowers the possibility of in-flight emergencies.

In order to ensure that any problems are found and fixed before they become major ones, anomaly detection has become a crucial tool for aircraft maintenance and operations. These algorithms gradually increase detecting skills by learning from fresh data, which raises the dependability and safety of aeronautical systems.

Through anomaly detection, non-destructive testing, and predictive maintenance, machine learning is transforming aircraft safety and quality assurance. These cutting-edge features enable businesses to keep an eye on the condition of its equipment, spot internal flaws, and spot anomalies instantly, guaranteeing maximum operational effectiveness and safety. The aerospace sector will be able to achieve strict safety criteria and provide dependable, high-performance systems as ML technologies continue to advance, making its contribution to improving aerospace safety even more crucial.

CHALLENGES AND ETHICAL CONSIDERATIONS

Although ML has many advantages in the aerospace industry, there are drawbacks and moral dilemmas when putting it into practice. To guarantee that ML adoption complies with safety, transparency, and social responsibility norms, these concerns need to be properly addressed. Three main issues are examined in this chapter: the ethical ramifications of automation in aerospace, explainability and trust, and data limits.

Data Challenges

The availability of diverse, high-quality datasets for training is a major factor in the efficacy of machine learning models. However, there are particular difficulties with data access and gathering in the aircraft sector.

Limited Historical Data for Model Training: Long life cycles and specialized construction characterize aerospace systems, such as flight control systems and aircraft engines. Because of this, there is a dearth of historical data accessible for ML model training, which hinders the models' ability to generalize well to new systems or designs. Furthermore, the amount of operational data that can be freely shared is limited by safety concerns, particularly in aerospace projects tied to defense.

High Costs of Data Generation and Experimentation: In the aerospace industry, creating new datasets frequently necessitates costly physical testing, such flight tests or wind tunnel experiments. Aerospace companies must rely significantly on simulations, which may not always convey the complexity of real-world conditions, in contrast to industries where experimentation can be done at little cost. The accuracy and resilience of ML models are impacted by this limitation.

Data Quality and Sensor Noise: In extreme environmental conditions, data from aircraft sensors may occasionally be noisy or lacking. This poses a problem for machine learning algorithms, which depend on clean, organized datasets to function properly. Preprocessing and imputation techniques are crucial for handling inconsistent or missing data, but they complicate the application of machine learning.

Collaboration throughout the aerospace sector will be necessary to address data difficulties in order to exchange anonymized data, build sophisticated models, and produce synthetic datasets. These initiatives will aid in improving the accuracy and adaptability of ML models to the particular requirements of aeronautical applications.

Explainability and Trust in ML

Building confidence in ML systems' decision-making is crucial in safety-critical sectors like aerospace. However, a lot of sophisticated machine learning models, especially those that rely on deep learning, function as "black boxes," which means that they don't reveal anything about how it arrive at its findings.

Deep Learning Models' Lack of Explainability: Predictive maintenance, anomaly detection, and flight control optimization are just a few of the complicated aerospace jobs that frequently employ deep neural networks (DNNs). Even though these models are capable of achieving great accuracy, engineers and regulators may find it difficult to comprehend how they arrive at specific forecasts or recommendations due to the complexity of internal workings.

Regulatory approval and certification challenges: Aerospace systems must pass stringent certification procedures to guarantee they adhere to safety and dependability requirements. However, when the underlying algorithms are not entirely explicable, certifying ML-based systems becomes problematic. Transparency is necessary for regulatory authorities to make sure that these systems operate dependably in all circumstances, and the absence of explainability is a major deterrent to approval.

Creating Solutions for Explainable AI (XAI): By offering insights into the decision-making process of ML models, explainable AI techniques seek to improve its interpretability. By providing reasons for important forecasts, like why an engine component is expected to fail, XAI may increase trust in the aerospace industry and empower engineers to confidently validate and act upon these predictions. To increase the transparency and reliability of ML models, methods such as SHAP (SHapley Additive exPlanations) and LIME (Local Interpretable Model-agnostic Explanations) are being investigated.

Performance and explainability must be balanced if machine learning is to be successfully implemented in the aerospace industry. Regulatory bodies and engineers need to work together to create frameworks that facilitate the use of trustworthy, comprehensible machine learning models.

Ethical Considerations

Significant ethical questions are brought up by the incorporation of ML-driven automation in the aerospace industry, especially in light of the possible societal effects on skilled labor and job displacement.

Job displacement and workforce transition: Robotic assembly and automated quality control are two examples of repetitive manufacturing and operational duties that are replaced by ML-powered automation, which increases productivity. Although productivity is increased, skilled personnel who were previously involved in these operations may be displaced as a result. Concerns about unemployment and the need to retrain workers for new positions are raised by the trend towards automation.

Reskilling and Upskilling: The aerospace sector must make investments in reskilling and upskilling employees as it embraces more machine learning-based technologies. This will guarantee that workers stay relevant in a technologically advanced workplace. These could include data science, programming, and ML model creation positions, as well as training programs that empower employees to operate and manage automated systems.

Efficiency and Social Responsibility: Aerospace businesses need to reconcile increasing efficiency with upholding social responsibility. Automation can cut costs and production times, but businesses must also think about how it will affect workers and make sure the switch to automated systems is handled morally. This might entail offering displaced workers alternate employment options and implementing automation in stages.

Bias and Fairness in ML Models: The possibility of bias in ML algorithms also raises ethical questions. Predictive maintenance plans and aircraft route optimization are two examples of how ML models may generate biassed choices that impact operations if training datasets are not representative. To preserve confidence and prevent unforeseen outcomes, it will be essential to make sure ML models are equitable and accountable.

To create regulations that address the moral ramifications of ML adoption, governments, regulators, and aerospace businesses will need to collaborate. Promoting openness, assisting with worker migrations, and making sure automation is used in a way that is socially responsible are all part of this.

FUTURE TRENDS AND OPPORTUNITIES

With developments in artificial intelligence (AI), ML, and cutting-edge technologies like quantum computing, the aerospace sector is poised for a new era. As these technologies advance, they have the potential to improve human-machine collaboration, unlock previously unheard-of design skills, and make it possible to solve difficult issues that were thought to be unsolvable. Three major themes influencing the aerospace industry's future are covered in this chapter: collaborative human-machine systems,

AI-driven design advances, and the possible influence of quantum computing on machine learning applications in the aerospace industry.

AI-Driven Design Innovations

The combination of AI, ML, and materials science will influence the future generation of spacecraft and airplanes, allowing for the development of novel designs with improved sustainability and performance.

Automated Conceptual Design and Simulation: AI-driven tools will automate early-stage design processes by rapidly generating and evaluating multiple design concepts based on performance criteria such as aerodynamics, weight, and fuel efficiency. Without the need for real prototypes, AI-based simulation platforms, for instance, may evaluate airflow over aircraft surfaces and recommend the best wing shapes.

New Materials and Smart Structures: AI-driven design tools will be enhanced by developments in materials science, such as self-healing materials and lightweight composites. Engineers will be able to investigate new material combinations that optimize strength while reducing weight with the aid of machine learning techniques. Aircraft constructions might also incorporate smart materials that react to its surroundings, enhancing durability and safety.

Electric and Sustainable Aviation: AI will be crucial to the advancement of hybrid propulsion systems and electric aircraft, among other sustainable aerospace technologies. In order to increase the range and efficiency of electric aircraft and hasten the shift to greener aviation solutions, machine learning models will help optimize battery management, energy consumption, and aerodynamics.

In the future, artificial intelligence (AI) will not only simplify conventional design procedures but also make it possible to develop entirely original ideas, including self-flying cars and modular spacecraft that can be altered for various purposes.

Collaborative Human-Machine Systems

Collaborative human-machine systems—where engineers and machine learning algorithms collaborate in real-time to create creative solutions—will become more and more important in aerospace design and manufacturing in the future.

Human-AI Co-Creation: ML systems will serve as intelligent assistants to enhance human creativity and decision-making, not to replace engineers. Engineers can concentrate on higher-level activities like honing concepts and weighing trade-offs by using generative design tools, which, for instance, suggest several options based on design limitations.

Real-Time Design Iterations: During the design phase, engineers will be able to communicate with ML models through collaborative platforms, modifying the models in real time in response to performance feedback. For instance, the ML model would instantly produce new design possibilities that satisfy the updated requirements if an engineer working on a spacecraft design changed factors pertaining to material strength.

Improved Decision-Making with Digital Twins: Using digital twins, which are virtual representations of real systems, can improve human-machine cooperation even further. These digital twins will be continuously updated by ML algorithms using real-time sensor data, allowing engineers to model different situations, forecast results, and make well-informed design choices.

By lowering the time and expense involved in conventional design and production processes, collaborative human-machine systems will enable aerospace engineers to push the limits of innovation.

Quantum Computing and ML in Aerospace

By resolving optimization issues that conventional computing is now unable to handle, quantum computing holds the potential to completely transform machine learning applications in the aerospace industry.

Exponential Computational Power: By using quantum bits, or qubits, to execute several operations at once, quantum computers provide exponential gains in processing power. With this skill, ML models will be able to handle large datasets more effectively and address issues like material discovery, flight path prediction, and aerodynamic optimization.

Optimization of Complex Systems: Many aerospace challenges, such as fuel consumption optimization, air traffic management, and satellite constellation design, involve highly complex optimization problems. The aerospace sector may become more efficient as a whole if quantum machine learning algorithms were able to solve these issues far more quickly than traditional methods.

Enhancing Aerospace Simulations and Cryptography: Fluid dynamics analysis and structural stress testing are two examples of aerospace simulations that will benefit from the speed and precision that quantum computing will bring. In order to protect aerospace communication networks and guarantee the security and privacy of data sent between aircraft, satellites, and ground systems, it will also be essential in the development of new cryptographic techniques.

Although the development of practical quantum computers is still ongoing, great strides are being made in the direction of achieving its potential. In the upcoming years, aerospace businesses will need to spend in research and create ML models that are suitable for the integration of quantum computing.

The combination of AI, ML, and emerging technologies like quantum computing will shape aerospace design and manufacturing in the future. Next-generation aircraft and spacecraft with improved performance, sustainability, and safety will be made possible by AI-driven design advancements. Engineers will be able to work more productively and creatively thanks to collaborative human-machine systems that promote real-time design revisions. As quantum computing develops, it will open up new avenues for resolving challenging aerospace problems, transforming the sector.

These patterns demonstrate the revolutionary possibilities of cutting-edge aerospace technologies. However, engineers, researchers, and technology providers will need to work together continuously to realize these prospects. In order to stay ahead in a field that is becoming more and more competitive, aerospace companies must also continue to be flexible and embrace new tools and techniques. To redefining what is feasible in aerospace, the confluence of AI, ML, and quantum computing will influence air and space travel for many years to come.

CONCLUSION

By increasing manufacturing accuracy, boosting operational safety, and streamlining design processes, ML has already started to transform the aerospace sector. ML provides cutting-edge solutions that lower costs, boost productivity, and improve reliability, ranging from generative design and predictive maintenance to smart manufacturing and anomaly detection. Aerospace operations become safer and more sustainable as a result of these developments, which also make it possible for quicker design iterations, improved quality control, and proactive maintenance.

To fully realize ML's promise, however, issues including scarce data availability, models that are difficult to explain, and ethical worries about automation must be resolved. Opportunities for even more innovation will present themselves as the aerospace industry adopts collaborative human-machine systems and investigates cutting-edge technologies like quantum computing. In order to keep ahead of industry demands and overcome existing constraints, it will be crucial to continue investing in machine learning research and development. With careful work, machine learning will continue to play a key role in determining the direction of aerospace, guaranteeing safer skies, more efficient production, and innovative design opportunities.

REFERENCES

Acuña, J. (2021). Evaluación limnológica y microbiológica de la Laguna del Quilotoa, provincia de Cotopaxi – Ecuador. *UNIVERSIDAD CENTRAL DEL ECUADOR*, *14*, 99–100.

Arista, R., Mas, F., Morales-Palma, D., Oliva, M., & Vallellano, C. (2022). The role of Industrial Resources in Reconfigurable Aerospace Production Systems: A Preliminary Literature Review. In *IFAC-PapersOnLine* (Vol. 55, pp. 2719–2724). Elsevier B.V., DOI: 10.1016/j.ifacol.2022.10.124

Bertoni, A., Hallstedt, S. I., Dasari, S. K., & Andersson, P. (2020). Integration of value and sustainability assessment in design space exploration by machine learning: An aerospace application. *Design Science*, *6*, e2. Advance online publication. DOI: 10.1017/dsj.2019.29

Brunton, S. L., Kutz, J. N., Manohar, K., Aravkin, A. Y., Morgansen, K., Klemisch, J., & McDonald, D. (2021). Data-driven aerospace engineering: Reframing the industry with machine learning. *AIAA Journal*, *59*(8), 2820–2847. DOI: 10.2514/1.J060131

Caliari, T., Ribeiro, L., Pietrobelli, C., & Vezzani, A. (2023). Global value chains and sectoral innovation systems: An analysis of the aerospace industry. *Structural Change and Economic Dynamics*, *65*, 36–48. DOI: 10.1016/j.strueco.2023.02.004

Dasari, K., Mekala, S., & Kaka, J. R. (2024). Evaluation of UDP-Based DDoS Attack Detection by Neural Network Classifier with Convex Optimization and Activation Functions. *Ingénierie des Systèmes d'Information*, *29*(3), 3. DOI: 10.18280/isi.290321

Dodin, P., Xiao, J., Adulyasak, Y., Alamdari, N. E., Gauthier, L., Grangier, P., Lemaitre, P., & Hamilton, W. L. (2023). Bombardier Aftermarket Demand Forecast with Machine Learning. *INFORMS Journal on Applied Analytics*, *53*(6), 425–445. DOI: 10.1287/inte.2023.1164

Dube, P., & Hiravennavar, S. (2020). Machine Learning Approach to Predict Aerodynamic Performance of Underhood and Underbody Drag Enablers. In SAE Technical Papers (Vol. 2020-April). SAE International. https://doi.org/DOI: 10.4271/2020-01-0684

Elham, A. (2015). Adjoint quasi-three-dimensional aerodynamic solver for multi-fidelity wing aerodynamic shape optimization. *Aerospace Science and Technology*, *41*, 241–249. DOI: 10.1016/j.ast.2014.12.024

Falcão e Cunha, N., Gan, T. S., Curcio, E., Amorim, P., Almada-Lobo, B., & Grunow, M. (2023). Robust supply chain design with suppliers as system integrators: An aerospace case study. *International Journal of Production Research*, *61*(15), 5244–5265. DOI: 10.1080/00207543.2022.2099769

Goksidan, H. T. (2019). Value Creation in Project Management: The Effect of Project Lifecycle Management Measures in an Aerospace Company. Strategica: Upscaling Digital. *Transformations in Business & Economics*, (July), 724–734.

Hartl, D. J., Mabe, J. H., Benafan, O., Coda, A., Conduit, B., Padan, R., & Doren, B. V. (2015). Standardization of shape memory alloy test methods toward certification of aerospace applications. *Smart Materials and Structures*, *24*(8), 082001. Advance online publication. DOI: 10.1088/0964-1726/24/8/082001

. Javaid, M., Haleem, A., Singh, R. P., Rab, S., & Suman, R. (2021, January 1). Significance of sensors for industry 4.0: Roles, capabilities, and applications. Sensors International. KeAi Communications Co. https://doi.org/.DOI: 10.1016/j.sintl.2021.100110

Jiang, Y., Tran, T. H., & Williams, L. (2023). Machine learning and mixed reality for smart aviation: Applications and challenges. *Journal of Air Transport Management*, *111*, 102437. Advance online publication. DOI: 10.1016/j.jairtraman.2023.102437

Lee, S. C. (2023). A Literary Survey on the Challenges of Adopting Lean in Aerospace Production and Aviation Maintenance. *Journal of Aviation Technology and Engineering*, *12*(1). Advance online publication. DOI: 10.7771/2159-6670.1243

Leonard, C., Prasetiyo, E. E., & Putra, I. R. (2023). Design and implementation of electrical system morphing wing flight control on prototype light aircraft. *International Journal of Power Electronics and Drive Systems*, *14*(2), 781–788. DOI: 10.11591/ijpeds.v14.i2.pp781-788

. Li, J., Du, X., & Martins, J. R. R. A. (2022, October 1). Machine learning in aerodynamic shape optimization. Progress in Aerospace Sciences. Elsevier Ltd. https://doi.org/.DOI: 10.1016/j.paerosci.2022.100849

. Ma, H., Huang, X., Hu, Z., Chen, Y., Qian, D., Deng, J., & Hua, L. (2023, August 1). Multi-objective production scheduling optimization and management control system of complex aerospace components: a review. International Journal of Advanced Manufacturing Technology. Springer Science and Business Media Deutschland GmbH. https://doi.org/.DOI: 10.1007/s00170-023-11707-4

Ni, Z. (2022). Reframe the Field of Aerospace Engineering Via Machine Learning: Application and Comparison. In Journal of Physics: Conference Series (Vol. 2386). Institute of Physics. https://doi.org/DOI: 10.1088/1742-6596/2386/1/012031

Nogueira, R. P. R., Melicio, R., Valério, D., & Santos, L. F. F. M. (2023). Learning Methods and Predictive Modeling to Identify Failure by Human Factors in the Aviation Industry. *Applied Sciences (Basel, Switzerland)*, *13*(6), 4069. Advance online publication. DOI: 10.3390/app13064069

Sun, K., Wang, W., Cheng, R., Liang, Y., Xie, H., Wang, J., & Zhang, M. (2024). Evolutionary generative design of supercritical airfoils: An automated approach driven by small data. *Complex & Intelligent Systems*, *10*(1), 1167–1183. DOI: 10.1007/s40747-023-01214-0

. Torbali, M. E., Zolotas, A., & Avdelidis, N. P. (2023, February 1). A State-of-the-Art Review of Non-Destructive Testing Image Fusion and Critical Insights on the Inspection of Aerospace Composites towards Sustainable Maintenance Repair Operations. Applied Sciences (Switzerland). MDPI. https://doi.org/.DOI: 10.3390/app13042732

Warey, A., Raul, V., Kaushik, S., Han, T., & Chakravarty, R. (2023). Generative Inverse Design of Aerodynamic Shapes Using Conditional Invertible Neural Networks. *Journal of Computing and Information Science in Engineering*, *23*(3), 031006. Advance online publication. DOI: 10.1115/1.4054715

Xue, T., Chen, L., Hu, J., & Zhou, Q. (2023). A variable-fidelity multi-objective optimization method for aerospace structural design optimization. *Engineering Optimization*, *55*(7), 1133–1148. DOI: 10.1080/0305215X.2022.2061964

Yang, Y., Wei, K., Kang, R., & Wang, S. (2018). Multi-objective chance constrained programming of spare parts based on uncertainty theory. *IEEE Access : Practical Innovations, Open Solutions*, *6*, 50049–50054. DOI: 10.1109/ACCESS.2018.2860252

Zhan, Z., & Li, H. (2021). A novel approach based on the elastoplastic fatigue damage and machine learning models for life prediction of aerospace alloy parts fabricated by additive manufacturing. *International Journal of Fatigue*, *145*, 106089. Advance online publication. DOI: 10.1016/j.ijfatigue.2020.106089

. Zhong, D., Xia, Z., Zhu, Y., & Duan, J. (2023, April 1). Overview of predictive maintenance based on digital twin technology. Heliyon. Elsevier Ltd. https://doi.org/.DOI: 10.1016/j.heliyon.2023.e14534

Chapter 4
Next-Generation Computer Vision for Autonomous Drone Navigation: Cutting-Edge Techniques and Real-World Applications

Chikesh Ranjan
https://orcid.org/0000-0003-1197-601X
National Institute of Technology, Rourkela, India

Jonnalagadda Srinivas
National Institute of Technology, Rourkela, India

P. S. Balaji
https://orcid.org/0000-0002-6364-4466
National Institute of Technology, Rourkela, India

Kaushik Kumar
https://orcid.org/0000-0002-4237-2836
Birla Institute of Technology, India

DOI: 10.4018/979-8-3693-7525-9.ch004

ABSTRACT

This chapter explores the cutting-edge advancements in computer vision technologies that are revolutionizing autonomous drone navigation. As drones are increasingly deployed in diverse applications such as mapping, surveillance, search and rescue, and delivery services, the need for precise and reliable navigation systems is paramount. The chapter begins with an overview of the fundamental principles of computer vision and its role in enabling autonomous navigation. The integration of machine learning algorithms, such as convolutional neural networks (CNNs) and deep learning frameworks, is discussed, highlighting their impact on enhancing the accuracy and efficiency of drone navigation. Furthermore, the chapter examines practical applications of these techniques, showcasing real-world scenarios where computer vision has significantly improved the performance of autonomous drones. Through this comprehensive examination, the chapter aims to provide valuable insights for researchers, engineers, and practitioners working to advance the field of autonomous drone navigation.

INTRODUCTION

Advancements in computer vision have revolutionized the field of autonomous drone navigation, allowing for more precise and efficient operations in dynamic environments. Drones can now navigate with greater autonomy by integrating key techniques like object detection, visual SLAM (Simultaneous Localization and Mapping), and deep learning-based obstacle avoidance (Arafat et al. 2023). These techniques enable drones to recognize and avoid obstacles, map environments in real-time, and adjust flight paths accordingly.

One of the most significant impacts is seen in agriculture, where drones can autonomously survey crops, and logistics, where they assist in package delivery. In defense, autonomous drones are used for surveillance and reconnaissance missions, offering enhanced safety and operational efficiency. Using AI-driven algorithms for real-time data processing enables drones to make intelligent decisions, further expanding their application in urban environments and complex terrains (Katkuri et al. 2024). As these technologies evolve, the future of autonomous drones promises even greater capabilities, transforming industries with enhanced precision and adaptability.

Computer vision techniques for autonomous drone navigation play a crucial role in enabling drones to operate independently in complex environments. As shown in Figure 1, the flow diagram presents the key computer vision techniques fundamental to autonomous drone navigation, each playing a crucial role in enabling the drone to perceive and respond to its surroundings.

Figure 1. Flow diagram of key computer vision techniques for autonomous drone navigation

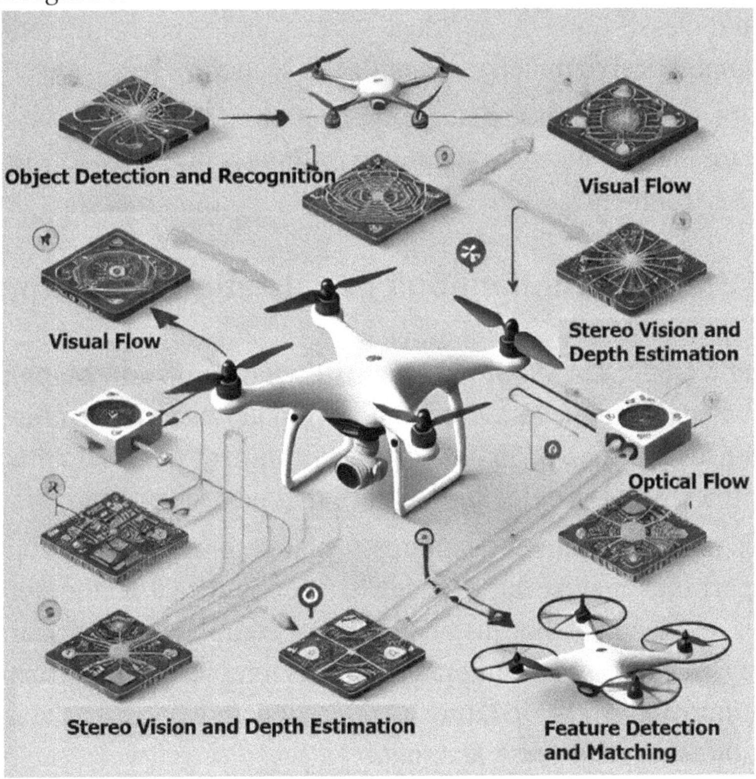

Key techniques include:

Object Detection and Recognition

Object detection and recognition are critical tasks in computer vision, enabling machines to identify and locate objects within images or video frames.

In object detection, algorithms are designed to detect the presence of objects in an image and determine their locations, typically by drawing bounding boxes around them. For example, a self-driving car might use object detection to identify pedestrians, vehicles, and traffic signs in its surroundings, helping it navigate safely (Bijjahalli, et al. 2020).

Object recognition, on the other hand, goes a step further by not only identifying the objects present but also classifying them into predefined categories. For instance, in an image of a park, an algorithm might recognize and classify objects as "dog," "tree," and "bench," allowing for a more comprehensive understanding of the scene.

Together, these processes enable applications ranging from surveillance systems to augmented reality, enhancing interaction between machines and the environment.

Visual SLAM (Simultaneous Localization and Mapping)

Object Visual SLAM (Simultaneous Localization and Mapping) is a sophisticated technique in robotics and computer vision that allows a system to construct a map of an unknown environment while simultaneously keeping track of its own location within that environment.

In this process, a camera captures visual data from the surroundings, while algorithms analyze these images to detect and track features or objects (Aburaya, et al. 2024). For instance, a robot equipped with visual SLAM might navigate through a cluttered room. As it moves, it identifies objects like furniture and walls, updating its internal map while also determining its position relative to these features.

An example of this in action is a drone flying indoors. Using visual SLAM, the drone can navigate through the space by recognizing landmarks such as doorways and windows, allowing it to avoid obstacles and create a detailed map of the interior (Upadhyay, et al. 2021). This capability is essential for applications like autonomous navigation, augmented reality, and robotic exploration. Here is the illustrative Figure 2 depicting Object Visual SLAM in action. It shows a robot navigating through a cluttered indoor environment while mapping its surroundings.

Figure 2. Object visual SLAM in action

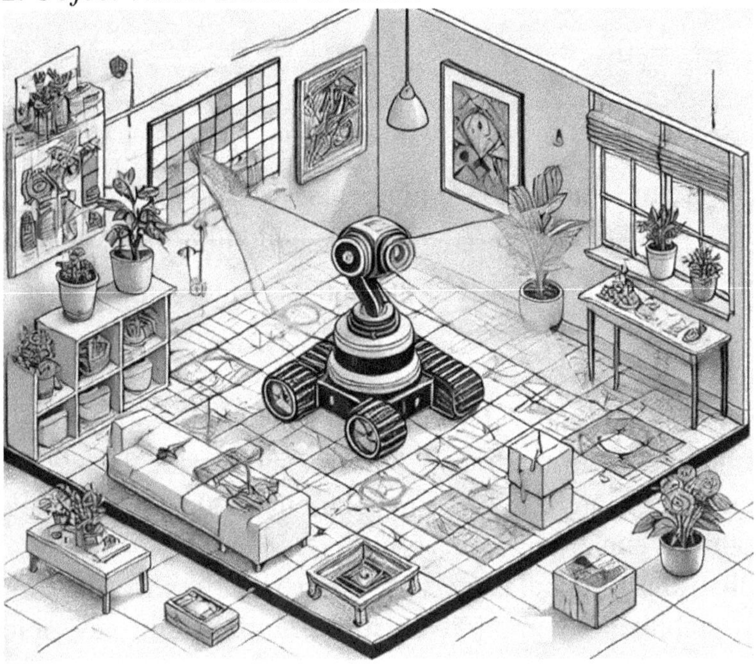

Optical Flow

Optical flow is a computer vision technique that estimates motion between two consecutive frames in a video sequence. It analyzes the apparent motion of objects and surfaces based on changes in brightness and pixel intensity over time. The key concepts include:

1. **Motion Estimation**: Optical flow calculates the motion vectors that describe how pixels move from one frame to another, helping to identify moving objects and their trajectories.

2. **Brightness Constancy Assumption**: It assumes that the brightness of a point in an image remains constant as it moves between frames. This allows for the estimation of motion based on changes in pixel positions.
3. **Optical Flow Equation**: The basic optical flow equation can be expressed as:

$$I_x u + I_y v + I_t = 0 \quad (1)$$

Where:

I_x and I_y are the spatial image gradients (derivatives for x and y).

I_t is the temporal image gradient (change in intensity over time).

u and v are the components of the optical flow vector (horizontal and vertical motion).

4. **Algorithms:** Common algorithms for calculating optical flow include the Lucas-Kanade method (which assumes a local neighbourhood for each pixel) and the Horn-Schunck method (which uses a global smoothness constraint).
5. **Applications**: Optical flow is used in various applications, including video stabilization, motion tracking, autonomous navigation, and activity recognition.

Deep Reinforcement Learning

Deep Reinforcement Learning (Deep RL) combines Reinforcement Learning (RL) and Deep Learning. It enables an agent to make decisions by interacting with an environment and learning through trial and error while using deep neural networks to approximate complex functions and policies (Senhaji-Mouhaddib, et al. 2024). Here's an overview of its key concepts:

Reinforcement Learning (RL) Basics

- **Agent and Environment**: In RL, an agent interacts with an environment to achieve a goal. At each step, the agent observes the current state of the environment, takes action and receives a reward based on the outcome.
- **States, Actions, and Rewards**:
    - o **State (S)**: The current situation or condition the agent is in.
    - o **Action (A)**: The choices the agent can make.
    - o **Reward (R)**: A feedback signal indicating the quality of the action taken, helping the agent learn over time.
- **Goal**: The agent aims to learn a policy, a mapping from states to actions, that maximizes the cumulative reward over time.

Deep Learning Integration

In complex environments with large state and action spaces, it's challenging to represent all possible states or actions using traditional tabular methods (Padmavathi, et al. 2024). Deep RL uses deep neural networks to approximate functions and make decisions efficiently:

- **Deep Q-Networks (DQN)**: A common Deep RL technique that uses neural networks to approximate the Q-value function, which represents the expected cumulative reward for taking an action in a given state.
- **Policy Gradient Methods**: Instead of learning the value of actions, policy gradient methods directly learn the policy (i.e., the action probabilities) using a neural network.

Key Components

- **Q-Function**: In **value-based methods** like DQN, the agent learns the expected reward for each action, using a Q-function Q (s.a) which is approximated by a neural network.

- **Policy Function**: In **policy-based methods**, the agent learns the probability distribution over actions given a state using a policy function π(s) also approximated by a deep network.
- **Exploration vs. Exploitation**: The agent must balance between exploring new actions (which may yield better rewards in the long run) and exploiting known actions that have previously provided good rewards.

Popular Algorithms

- **Deep Q-Network (DQN)**: Combines Q-learning with deep learning. It uses experience replay and target networks to stabilize training.
- **Proximal Policy Optimization (PPO)**: A policy gradient method that updates the policy more cautiously to prevent drastic changes during training, improving stability.
- **Actor-Critic Methods**: These involve two models: an actor (which selects actions) and a critic (which evaluates actions), combining the benefits of both value-based and policy-based methods.

Applications

Deep RL has been successfully applied in a variety of complex domains:

- **Game Playing**: Notably, DeepMind's AlphaGo used deep RL to defeat human champions in Go.
- **Robotics**: Autonomous robots use Deep RL for tasks like grasping objects and navigation.
- **Autonomous Vehicles**: Cars can learn to navigate safely through environments.
- **Natural Language Processing**: Dialogue systems and chatbots can optimize conversation strategies using Deep RL.

Challenges

- **Sample Efficiency**: Training requires a vast amount of data and time, as agents learn through trial and error.

- **Stability and Convergence**: Deep RL models can be unstable and may not always converge due to the high dimensionality of state and action spaces.
- **Exploration in Large State Spaces**: Ensuring efficient exploration in complex environments remains a challenge.

Deep Reinforcement Learning combines the strengths of reinforcement learning with the powerful function approximation of deep learning, enabling agents to solve highly complex tasks in dynamic environments.

Deep Reinforcement Learning

Deep Reinforcement Learning (Deep RL) is a method that combines reinforcement learning (RL) with deep learning to allow agents to learn optimal actions in complex environments. Here's a breakdown:

1. **Reinforcement Learning (RL)**: In RL, an agent interacts with an environment by taking actions, receiving feedback (rewards or penalties), and adjusting its behavior to maximize cumulative rewards over time. The goal is to learn a policy (set of actions) that maximizes the long-term reward.
2. **Deep Learning**: Deep neural networks are used to approximate functions (like the value function or policy) when the state and action spaces are too large for traditional RL methods. Neural networks are especially useful for handling high-dimensional input data, such as images or complex sensor readings.
3. **Key Concepts**:
    - **State**: A representation of the environment at a specific moment.
    - **Action**: The decision the agent makes in response to the state.
    - **Reward**: The feedback signal indicating the success of an action.
    - **Policy**: A function that maps states to actions (can be deterministic or stochastic).
    - **Value Function**: Estimates how good a specific state (or state-action pair) is in terms of expected future rewards.

4. **Deep Q-Networks (DQN)**: One of the foundational methods in Deep RL, where a neural network is used to approximate the Q-value function. The Q-value represents the expected reward for taking a certain action in a given state. DQN was famously used by DeepMind to surpass human-level performance in Atari games.
5. **Policy Gradient Methods**: These methods, such as Proximal Policy Optimization (PPO) and Actor-Critic algorithms, directly optimize the policy (rather than approximating value functions like in DQN). These methods are effective for continuous action spaces.
6. **Exploration vs. Exploitation**: The agent must balance between exploring new actions to discover better policies (exploration) and choosing the best-known actions to maximize reward (exploitation).
7. **Drones and Deep RL**: Drones equipped with this technology can learn from their environment through trial and error. Reinforcement learning models enable drones to make decisions on the fly, improving their ability to avoid obstacles and navigate complex spaces autonomously.
8. **Applications**: Deep RL has been used in gaming (e.g., AlphaGo), robotics, autonomous driving, resource management, and more.

Stereo Vision and Depth Estimation

Stereo Vision and Depth Estimation refer to techniques that enable machines or systems to perceive the 3D structure of their environment by using two or more cameras (Konovalenko, et al. 2018). Here's how they work:

1. **Stereo Vision**: This is a method inspired by human binocular vision. Two cameras, placed at different viewpoints (like two eyes), capture images of the same scene. By comparing the differences between the two images (known as disparity), the system can infer depth information, or how far away objects are from the cameras.
2. **Depth Estimation**: This is the process of calculating the distance of objects from the camera(s). In stereo vision, depth estimation is done by finding corresponding points in the two images and measuring the disparity (difference in pixel position). Objects closer to the camera have greater disparity, while farther objects have smaller disparity.

3. **Key Steps in Stereo Vision**:
   - o **Image Capture**: Two or more cameras capture images from different angles.
   - o **Feature Matching**: The system finds corresponding points (features) in both images.
   - o **Disparity Calculation**: The difference in position of these points between the two images is used to calculate disparity.
   - o **Depth Calculation**: Using the disparity values and knowing the distance between the cameras (baseline), the system calculates the depth of each point in the scene.
4. **Drones and Stereo Vision**: By using multiple cameras or sensors, drones can create 3D representations of their surroundings, allowing them to better understand distances to objects. This capability is crucial for collision avoidance and navigation in cluttered spaces, where precise depth perception is needed to maneuver safely.
5. **Applications**: Stereo vision and depth estimation are widely used in robotics, autonomous vehicles, drones, augmented reality, and 3D mapping.

By using multiple cameras or sensors, drones can create 3D representations of their surroundings, allowing them to better understand distances to objects, crucial for collision avoidance and navigation in cluttered spaces.

Feature Detection and Matching

Feature Detection and Matching are essential techniques in computer vision used to identify and match key points or patterns in images (Rodriguez, et al. 2024). These methods help systems recognize objects, track movements, and understand the structure of the environment. Here's how they work:

1. **Feature Detection**: In an image, certain points or regions stand out due to their distinct properties (e.g., corners, edges, or textures). These points, called features, are identifiable and consistent across different views or lighting conditions. The goal of feature detection is to find

these key points in an image, which can later be used for further analysis like tracking or 3D reconstruction.

2. **Common Feature Detection Algorithms**:
   - **SIFT (Scale-Invariant Feature Transform)**: SIFT detects distinctive points in an image and generates descriptors that are invariant to scaling, rotation, and even illumination changes. It helps systems recognize and match features across different perspectives.
   - **ORB (Oriented FAST and Rotated BRIEF)**: ORB is a faster, more efficient alternative to SIFT, designed for real-time applications. It uses the FAST corner detector to find key points and the BRIEF descriptor to represent the appearance of the detected features. ORB is robust to rotation and is widely used in resource-constrained systems like drones.
3. **Feature Matching**: Once key points are detected, the next step is to match corresponding features between different images or frames. This is typically done by comparing the feature descriptors. The goal is to find similar points in two images that represent the same physical point in the environment.
4. **Drones and Feature Detection**: Algorithms like SIFT and ORB help drones detect and track features in the environment, aiding in localization (understanding the drone's position relative to its surroundings), mapping (building a map of the environment), and scene understanding (recognizing objects or obstacles). These capabilities are essential for autonomous navigation, as the drone needs to understand and interact with its surroundings in real-time.
5. **Applications**: Feature detection and matching are crucial for tasks like Simultaneous Localization and Mapping (SLAM), 3D reconstruction, object recognition, and motion tracking. These tasks are fundamental to robotics, augmented reality, and autonomous vehicles.

Computer Vision for Drone Navigation

Computer Vision for Drone Navigation enables drones to perceive and interpret their environment, allowing for autonomous navigation. By utilizing visual data from cameras and sensors, drones can make real-time decisions regarding path planning, obstacle avoidance, and interaction with objects (Tong, et al. 2023). This technology simulates human vision to assist drones in navigating complex environments.

1. **Perception**: Drones are equipped with cameras or vision sensors that capture images and videos of their surroundings. Computer vision algorithms process this visual data to extract critical information, such as detecting obstacles, recognizing landmarks, and identifying navigable paths.
2. **Key Techniques in Computer Vision**:
   - **Stereo Vision and Depth Estimation**: By using multiple cameras, drones can estimate depth and create 3D maps of their environment by comparing different views of the same scene. This capability allows drones to understand distances to objects, crucial for obstacle avoidance and maneuvering in cluttered spaces.
   - **Optical Flow**: Optical flow tracks the movement of objects by analyzing changes in pixel intensities between consecutive frames, helping drones understand motion and estimate their own velocity relative to their surroundings.
   - **Feature Detection and Matching**: Algorithms such as SIFT (Scale-Invariant Feature Transform) and ORB (Oriented FAST and Rotated BRIEF) allow drones to detect and track distinctive features in their environment, aiding in localization (identifying their position) and mapping (building a map of the area). This is essential for autonomous navigation in unknown environments.
3. **Simultaneous Localization and Mapping (SLAM)**: SLAM is a process by which drones build a map of their environment while simultaneously determining their own position within it. Using visual data and additional sensors (like IMUs), drones can navigate through unknown spaces without relying on GPS.
4. **Obstacle Detection and Avoidance**: Drones use real-time computer vision to detect and avoid obstacles, ensuring safe navigation. Techniques such as stereo vision, depth estimation, and optical flow allow drones to assess distances and safely maneuver around potential collisions.

5. **Autonomous Navigation**: Computer vision algorithms enable drones to make real-time decisions about flight paths, speed, and direction, even in complex environments like urban areas or indoor spaces. This allows drones to fly autonomously without human intervention.
6. **Integration with AI**: By incorporating advanced AI models and deep learning, drones can further enhance their scene understanding, recognize objects, and predict the movement of dynamic obstacles such as cars or pedestrians, leading to more intelligent navigation (Qian, et al. 2024).
7. **Applications**:
    - **Search and Rescue**: Drones equipped with computer vision can autonomously navigate through challenging terrain or disaster zones, identifying visual cues to locate survivors.
    - **Agriculture**: In agriculture, drones use computer vision to monitor crop health and identify areas that require attention based on aerial imagery.
    - **Delivery Services**: Autonomous delivery drones utilize vision systems to safely navigate to delivery points, avoiding obstacles and ensuring efficient route planning.

Computer Vision-Based Landing for UAVs

Computer vision-based landing for Unmanned Aerial Vehicles (UAVs) is an advanced technique that enhances autonomous flight capabilities by enabling precise and safe landing without relying on external navigation aids like GPS. In this approach, UAVs utilize onboard cameras and computer vision algorithms to detect and analyze visual cues from the environment, such as landmarks, markers, or designated landing zones. These cues are processed in real-time to guide the UAV during its descent and final approach, ensuring an accurate landing even in challenging conditions or GPS-denied environments (Thangamani, et al. 2024).

The computer vision system typically employs image processing techniques such as object detection, pattern recognition, and feature extraction to identify the landing target. Machine learning or deep learning models are often integrated to improve the system's ability to recognize complex patterns and adapt to varying lighting conditions, terrain textures, and environmental factors (Xin, et al. 2022). For example, convolutional neural networks (CNNs) can be used to detect and classify landing markers or identify specific textures on the ground.

This method of landing is particularly valuable for UAV operations in environments where GPS signals are weak, unreliable, or unavailable, such as urban areas, forests, or indoor spaces. It also enhances the versatility of UAVs for applications like package delivery, search and rescue missions, and agricultural monitoring, where accurate landing in specific locations is critical.

By leveraging computer vision, UAVs can autonomously adjust their landing path in real-time, ensuring safety and precision, which reduces the risk of crashes or landing in unintended areas. As research in computer vision and autonomous systems advances, these techniques will continue to improve, contributing to the broader adoption of UAVs in both civilian and military applications (Hanover, et al. 2022).

Computer Vision-based Surveillance and Tracking for UAVs

Computer vision-based surveillance and tracking for UAVs is an emerging area that utilizes advanced imaging and pattern recognition to enable real-time monitoring and navigation (Zhong, et al. 2022). Here are some key components and approaches:

- **Object Detection and Tracking:** UAVs use computer vision algorithms to detect and track objects of interest, distinguishing between humans, vehicles, and other objects. Techniques such as YOLO (You Only Look Once), Faster R-CNN, and SSD (Single Shot Detector) are common for real-time object detection.
- **Image and Video Processing:** Using onboard cameras, UAVs process frames to identify movements, anomalies, or suspicious activity. Image stabilization, segmentation, and background subtraction are essential for handling real-time video feeds, especially in varying light and weather conditions.

- **Feature Extraction and Recognition:** UAVs analyze features like shape, color, and texture for accurate identification of objects and targets. Machine learning algorithms, especially convolutional neural networks (CNNs), are crucial for feature extraction in complex scenarios.
- **3D Mapping and Localization:** UAVs equipped with LiDAR or stereo cameras use 3D reconstruction and SLAM (Simultaneous Localization and Mapping) to create real-time maps, aiding in navigation and precise object tracking.
- **Multi-UAV Coordination:** Coordinated UAV systems enhance surveillance coverage. Techniques like cooperative tracking and decentralized control enable multiple UAVs to share data and collaborate in complex environments.
- **Human Activity Recognition:** In security scenarios, UAVs use human activity recognition models to assess movements, detecting behaviors such as running, loitering, or suspicious gestures to trigger alerts.

Challenges and Opportunities

In the field of autonomous drone navigation, computer vision plays a critical role but brings with it distinct challenges and opportunities:

Challenges

- Real-Time Processing Requirements

Autonomous navigation demands immediate decision-making. Processing high-resolution visual data in real-time, often on limited onboard hardware, is computationally intense, limiting drones' ability to make rapid and accurate responses (Badrloo, et al. 2022).

- Environmental Variability

Drones encounter diverse conditions (e.g., weather changes, lighting variations, obstacles) that challenge visual sensors and require robust algorithms to handle sudden shifts without compromising safety or accuracy.

- Power and Battery Constraints

High computational workloads for computer vision algorithms quickly drain drone batteries, reducing flight time and limiting mission duration. Developing energy-efficient processing methods is essential.

- Localization and Mapping in GPS-Denied Environments

GPS is unreliable in areas like dense forests or urban canyons, necessitating alternative visual-based localization and mapping techniques, such as Visual SLAM (Simultaneous Localization and Mapping), which adds computational complexity(Be co, et al. 2024).

- Complex Object Detection and Tracking

Identifying and tracking various objects (e.g., people, vehicles) can be difficult in dynamic or cluttered environments. Achieving reliable tracking across different angles, speeds, and occlusions is still challenging (Dwivedi, et al. 2023).

- Data Privacy and Security

Using visual data for surveillance in public areas raises privacy concerns. Securing sensitive data and ensuring ethical deployment while maintaining public trust is a key obstacle.

Opportunities

- Advancements in AI and Deep Learning

Improved algorithms, like convolutional neural networks (CNNs) and transformers, enhance object recognition and depth estimation capabilities. These innovations enable drones to better understand complex scenes and respond intelligently.

- Edge Computing and Specialized Hardware

Edge AI devices and dedicated GPUs/TPUs (Graphics/TPU Units) allow faster, more power-efficient processing directly on drones, reducing latency and enabling real-time analytics on lightweight hardware (Rangel, et al. 2024)..

- Enhanced 3D Mapping and Localization

Computer vision advances in 3D mapping and depth sensing, such as LiDAR integration and stereo vision, enable drones to navigate and interact within complex 3D spaces, opening possibilities for autonomous indoor and outdoor applications.

- Multi-Sensor Fusion

Integrating data from multiple sensors (e.g., IMUs, LiDAR, thermal cameras) improves reliability and accuracy in various environments. Multi-sensor fusion helps drones make better-informed decisions and adapt to environmental changes.

- Collaboration and Swarm Intelligence

Swarm technology allows multiple drones to work together in mapping and surveillance. Through shared visual data, drones can cover larger areas, navigate collaboratively, and execute complex tasks that would be challenging for a single UAV.

- Applications in Diverse Sectors

Autonomous drone navigation offers solutions for delivery, search and rescue, agriculture, inspection, and disaster response. These applications highlight the potential for computer vision-driven drones to expand into new domains and contribute to productivity and safety.

CONCLUSION

The advancements in computer vision for autonomous drone navigation have revolutionized the capabilities and applications of UAVs across various sectors. This chapter explored how cutting-edge techniques, including real-time object detection, 3D mapping, multi-sensor fusion, and deep learning, empower drones to perceive, interpret, and navigate complex environments autonomously. Despite the considerable challenges, such as processing demands, environmental variability, and data privacy, the continuous evolution of hardware and algorithms is driving innovative solutions. These advancements unlock significant opportunities, enabling drones to operate reliably in challenging conditions, collaborate in swarms, and perform diverse tasks with greater accuracy and autonomy. Applications in areas such as surveillance, disaster response, inspection, agriculture, and logistics highlight the transformative potential of computer vision in extending drone utility. As technology progresses, integrating more efficient processing, improved AI models, and robust multi-sensor systems will be key to achieving safe, intelligent, and versatile UAV navigation. These developments lay a strong foundation for future research and industrial deployment, positioning autonomous drones as essential tools in a wide range of real-world applications.

FUTURE SCOPE

The future of computer vision in autonomous drone navigation promises transformative advancements, with ongoing research paving the way for more intelligent, resilient, and versatile UAV systems. Key areas of development include the integration of advanced AI-driven algorithms, such as generative and self-supervised learning, which can enhance drones' ability to adapt to unfamiliar environments and tasks without extensive pre-training. Further, miniaturized edge computing hardware and energy-efficient processing units will allow for faster, real-time decision-making onboard, addressing power constraints and enabling longer flight durations.

Multi-modal sensor fusion, combining vision with LiDAR, radar, and thermal imaging, will greatly improve situational awareness and navigation accuracy, especially in GPS-denied or visually challenging environments. Swarm intelligence and collaborative UAV networks represent another frontier, enabling multiple drones to coordinate tasks, share data, and cover larger areas effectively. Moreover, robust ethical frameworks and data privacy protections will be crucial as autonomous drones become more integrated into public and commercial spaces.

In applications, the future will see drones playing a significant role in automated delivery, infrastructure inspection, precision agriculture, and emergency response. These advancements will not only expand the practical use cases for autonomous UAVs but also drive adoption across industries, ultimately contributing to safer, more efficient, and adaptable drone operations in real-world settings.

REFERENCES

Aburaya, A., Selamat, H., & Muslim, M. T. (2024). Review of vision-based reinforcement learning for drone navigation. *International Journal of Intelligent Robotics and Applications*, *8*(4), 1–19. DOI: 10.1007/s41315-024-00356-9

Arafat, M. Y., Alam, M. M., & Moh, S. (2023). Vision-based navigation techniques for unmanned aerial vehicles: Review and challenges. *Drones (Basel)*, *7*(2), 89. DOI: 10.3390/drones7020089

Badrloo, S., Varshosaz, M., Pirasteh, S., & Li, J. (2022). Image-based obstacle detection methods for the safe navigation of unmanned vehicles: A review. *Remote Sensing (Basel)*, *14*(15), 3824. DOI: 10.3390/rs14153824

Be co, D., Pârvu, P. V., & Ciudin, S. (2024). Autonomous navigation for the Martian Inspection Drone. *Acta Astronautica*, *216*, 55–63. DOI: 10.1016/j.actaastro.2023.12.052

Bijjahalli, S., Sabatini, R., & Gardi, A. (2020). Advances in intelligent and autonomous navigation systems for small UAS. *Progress in Aerospace Sciences*, *115*, 100617. DOI: 10.1016/j.paerosci.2020.100617

Dwivedi, K., Govindarajan, P., Srinivasan, D., Keerthi Sanjana, A., Selvanambi, R., & Karuppiah, M. (2023). Intelligent Autonomous Drones in Industry 4.0. In Sarveshwaran, V., Chen, J. I. Z., & Pelusi, D. (Eds.), *Artificial Intelligence and Cyber Security in Industry 4.0. Advanced Technologies and Societal Change*. Springer., DOI: 10.1007/978-981-99-2115-7_6

Hanover, D., Loquercio, A., Bauersfeld, L., Romero, A., Penicka, R., Song, Y., Cioffi, G., Kaufmann, E., & Scaramuzza, D. (2024). Autonomous drone racing: A survey. *IEEE Transactions on Robotics*, *20*, 3044–3067. DOI: 10.1109/TRO.2024.3400838

Katkuri, A. V. R., Madan, H., Khatri, N., Abdul-Qawy, A. S. H., & Patnaik, K. S. (2024). Autonomous UAV Navigation using Deep Learning-Based Computer Vision Frameworks: A Systematic Literature Review. *Array (New York, N.Y.), 100361*, 100361. Advance online publication. DOI: 10.1016/j.array.2024.100361

Konovalenko, I., Kuznetsova, E., Miller, A., Miller, B., Popov, A., Shepelev, D., & Stepanyan, K. (2018). New approaches to the integration of navigation systems for autonomous unmanned vehicles (UAV). *Sensors (Basel), 18*(9), 3010. DOI: 10.3390/s18093010 PMID: 30205568

Padmavathi, A., & Vyasasreevedansh, B. (2024) Smart Surveillence Drone with Navigation System, *2024 IEEE International Conference on Computing, Power and Communication Technologies (IC2PCT)*, Greater Noida, India, *5*, 1452-1458, https://doi.org/DOI: 10.1109/IC2PCT60090.2024.10486563

Qian, B., Al Said, N., & Dong, B. (2024). New technologies for UAV navigation with real-time pattern recognition. *Ain Shams Engineering Journal, 15*(3), 102480. DOI: 10.1016/j.asej.2023.102480

Rangel, P., Tardif, S., Mehrubeoglu, M., John, E. S., Whaley, P., Salas, M., . . . Torres, M. (2024, June). IoT-enabled unmanned traffic management system with dynamic vision-based drone detection for sense and avoid coordination. In *Real-Time Image Processing and Deep Learning 2024* Orlando, Florida, US, SPIE *13034*, 99-111 https://doi.org/DOI: 10.1117/12.3014009

Rodriguez, A. A., Shekaramiz, M., & Masoum, M. A. (2024). Computer Vision-Based Path Planning with Indoor Low-Cost Autonomous Drones: An Educational Surrogate Project for Autonomous Wind Farm Navigation. *Drones (Basel), 8*(4), 154. DOI: 10.3390/drones8040154

Senhaji-Mouhaddib, I., Bouzakri, S., & Lagrat, I. (2024, May). Advanced Autonomous Navigation Technologies for UAVs: Challenges and Opportunities," *2024 4th International Conference on Innovative Research in Applied Science, Engineering and Technology (IRASET)*, FEZ, Morocco, 1-7, https://doi.org/DOI: 10.1109/IRASET60544.2024.10548645

Thangamani, R., Suguna, R. K., & Kamalam, G. K. (2024). Drones and Autonomous Robotics Incorporating Computational Intelligence. In: Prakash, K. B., Peddapelli, S. K., Tam, I. C. K., Woo, W. L., Jain, V. (eds) *Computational Intelligent Techniques in Mechatronics*, 243-296, John Wiley & Sons, Inc., Hoboken, NJ, US and Scrivener Publishing, Beverly, MA, US DOI: 10.1002/9781394175437.ch8

Tong, P., Yang, X., Yang, Y., Liu, W., & Wu, P. (2023). Multi-UAV collaborative absolute vision positioning and navigation: A survey and discussion. *Drones (Basel)*, *7*(4), 261. DOI: 10.3390/drones7040261

Upadhyay, J., Rawat, A., & Deb, D. (2021). Multiple drone navigation and formation using selective target tracking-based computer vision. *Electronics (Basel)*, *10*(17), 2125. DOI: 10.3390/electronics10172125

Xin, L., Tang, Z., Gai, W., & Liu, H. (2022). Vision-based autonomous landing for the UAV: A review. *Aerospace (Basel, Switzerland)*, *9*(11), 634. DOI: 10.3390/aerospace9110634

Zhong, J., Li, M., Chen, Y., Wei, Z., Yang, F., & Shen, H. (2024). A safer vision-based autonomous planning system for quadrotor UAVs with dynamic obstacle trajectory prediction and its application with LLMs. *IEEE/CVF Winter Conference on Applications of Computer Vision Workshops (WACVW)*, Waikoloa, HI, USA, 920-929, https://doi.org/DOI: 10.1109/WACVW60836.2024.00131

Chapter 5
Multi-Scene Recognition in Single Aerial Images Using CNN

G. Ananthi
https://orcid.org/0000-0003-3227-2134
Mepco Schlenk Engineering College, India

S. Mehala Shevani
Mepco Schlenk Engineering College, India

P. Priyadharshini Devi
Mepco Schlenk Engineering College, India

ABSTRACT

Scene recognition from aerial images is a primary application in utilizing high resolution satellite images. In the recent past, many studies were carried out in this field to classify an image to one scene category. But the real scenarios are different, they often contain more than one scene. Therefore, we explore an approach to recognize multiple scenes from a single aerial image. To do this, three different datasets namely UCM dataset, AID dataset and MAI dataset consisting of 2100, 10000, 3923 images respectively were used. Out of the three datasets, UCM and AID dataset

DOI: 10.4018/979-8-3693-7525-9.ch005

are Single scene datasets and the MAI dataset is a multi-scene dataset. These datasets are utilized in two configurations namely UCM2MAI and AID2MAI. Prototype based memory network using different CNN baseline models is used as the backbone. First, the single scene datasets were used to extensively train different baseline models and store them as prototypes in the external memory. Experiments were conducted to analyse the performance of different baseline models.

INTRODUCTION

In recent years, the scopes and applications of remote sensing techniques have been phenomenal, ranging from urban cartography to traffic monitoring, terrain surface analysis to ecological scrutiny, and determining the damage to the landscape caused by natural calamities such as earthquakes, tsunamis, hurricanes, and so on. These are some of the notable developments aided by aerial image classification. Because of the vast and vital scopes of aerial image classification, it has become a hot topic among researchers.

To begin, aerial image classification refers to providing images with labels associated with the scene visible on the images based on a single semantic category. Because satellites capture aerial images from great distances, they contain not only a single scene but also many objects in the landscape. This macroscopic view of single object classification may sometimes result in ignoring the major issue to be addressed by rendering the model infeasible in providing a comprehensive view of scenes in images for mission-critical applications.

A large number of algorithms have been proposed in the literature to identify each and every pixel of an image, localize objects with bounding boxes, and recognize the ground truth. However, these methods often require precise annotations, high computational power, and access to high-quality datasets—resources that are not always readily available. Furthermore, acquiring and curating relevant remote sensing datasets can be expensive and time-consuming due to the need for manual annotation and domain expertise.

Multilabel image classification comes into play here. Unlike traditional single-label classification, which assumes each image corresponds to only one category, multilabel classification assigns multiple semantic labels to an image, acknowledging the complexity and richness of real-world aerial scenes. Image classification with multiple labels can give a thorough overview of all image scenes that can be drawn. Figure 1 depicts the distinction between a single scene classifier and a multilabel classifier. In Figure 1, single-scene images predominantly feature one dominant class or object category. The assumption is that the entire image can be associated with a single semantic label that best describes its overall content. For example, an aerial image entirely covered by a dense forest would be labeled as "forest," or an image of a water body would be labeled as "lake" or "river." Multi-scene images, by contrast, contain multiple distinct object categories or land cover types within the same frame. These images reflect the real-world complexity captured by aerial and satellite sensors. For example, a single aerial photo of a suburban area might include houses, roads, vehicles, vegetation, and water bodies—each of which carries semantic importance.

The transition from single-label to multilabel classification in aerial images marks a paradigm shift. This approach is particularly valuable in heterogeneous environments such as urban landscapes, agricultural fields, or disaster zones, where multiple land cover types or events may co-occur. For instance, a single image might simultaneously depict water bodies, residential buildings, roads, and vegetation—all of which are critical for accurate scene interpretation.

The advancements in deep learning, especially Convolutional Neural Networks (CNNs), Transformer-based models, and hybrid architectures, have significantly improved the performance of multilabel classification tasks. These models can automatically learn hierarchical features from raw images, making them well-suited for the high-dimensional nature of aerial data. Moreover, attention mechanisms and context-aware learning approaches further enhance the model's ability to focus on relevant parts of an image for each label.

In addition, the integration of ancillary data such as metadata, digital elevation models (DEMs), and multi-temporal imagery with aerial photos has opened up new possibilities for more accurate and context-rich classification. Research is now increasingly focused on designing models that not only provide accurate predictions but also offer interpretability and robustness, which are crucial for deployment in operational remote sensing systems.

In conclusion, the growing complexity of real-world environments captured in aerial imagery has necessitated a shift towards more comprehensive and context-aware classification methods. Multilabel aerial image classification emerges as a powerful tool that aligns with the needs of diverse remote sensing applications, paving the way for more intelligent, scalable, and actionable geospatial analysis systems.

Figure 1. An example illustration on how the prototype-based memory network learns & memorizes each scene and stores in external memory and when a query muti-scene image is provided how the attentional memory retrieval module retrieves the scenes in the image

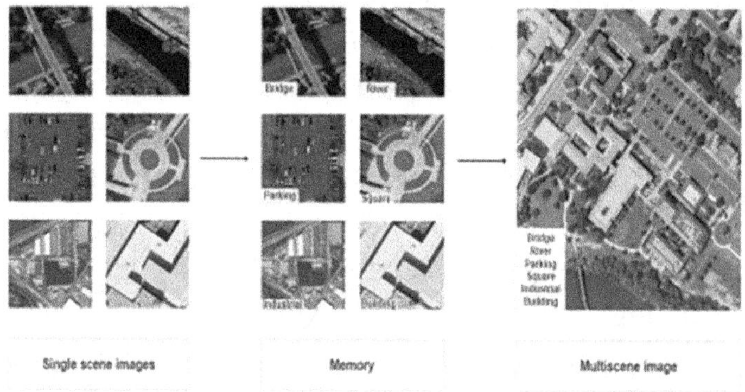

BACKGROUND AND LITERATURE REVIEW

There are fewer literatures on classifying multiple labels in a single image because of the following major challenges associated with multi-label classification:

1. Extracting multiple sematic categories from raw aerial images.

2. The approach involves leveraging label correlations to predict multiple scene labels from a single aerial image.

The former is critical and serves as the foundation for the proposed work, but it is difficult because aerial images are frequently high-resolution images and contain spatial contextual information. The baseline approaches underperform as their recognizing scope is limited to manually crafted semantic features, and they eventually fail to achieve high classification performance. As a result, the need for a higher-level feature extractor is indispensable.

The latter, on the other hand, implies modelling label-to-label correlations, which is important in the identification of multiple objects in aerial images. Single object classification, on the other hand, requires only image-label relevance. For example, the presence of cars indicates a high likelihood of the co-occurrence of roads, pavements, or parking lots. Similarly, the presence of a boat strongly suggests the presence of a body of water. Such studies are extremely rare in the literature. So, after considering the potential challenges, the solution is expected to be reliable in learning high-level feature representations, with label-to-label correlation being highly relevant.

Memory Networks are a type of neural network that have the ability to read, write and retrieve information from an external memory module. These memory networks were introduced (Sukhbaatar et al. 2015). They proposed a Memory-augmented network for object recognition tasks. The authors used a Long Short-Term Memory (LSTM) network to store and retrieve information from the memory module. The Memory Networks allowed the network to remember rare events and objects, improving the overall performance of the system. Eslami et al. (2016) used a Memory-augmented network to generate images of complex scenes. The authors used the memory module to store and retrieve information about the objects in the scene, allowing the system to generate more accurate and detailed images. Snell et al. (2017) introduced Prototypical Networks and used them for few-shot image classification tasks. Few-shot learning is a type of machine learning where the goal is to learn from a small number of labeled examples. The authors used the distance metric module to compute the distance between the input image and the prototypes, and the classification module to assign the input image to the class with the closest prototype.

Vinyals et al. (2016) extended the Prototypical Networks to generate imaginary samples to improve performance in few-shot image classification tasks. The authors generated these imaginary samples by interpolating between the prototypes of different classes. Hariharan and Girshick (2017) introduced Matching Networks, which are the variant of Prototypical Networks that use attention mechanisms to weight the importance of different parts of the input data. The authors used Matching Networks for one-shot image classification tasks. Sun et al. (2020) introduced Embedded Class Models, which are a variant of Prototypical Networks that learn a class-level representation of the data. The authors used Embedded Class Models for few-shot image classification tasks and achieved state-of-the-art performance on several benchmark datasets.

Zhou et al. (2018) showed the effectiveness of deep features from CNNs for scene classification but did not handle multi-label outputs. They demonstrated the effectiveness of deep convolutional features for scene classification, particularly using the Places dataset, but it did not address multi-label classification (where an image can belong to multiple scene categories).

Cheng et al. (2017) adapted pre-trained CNNs for aerial scenes but needed task-specific fine-tuning. They evaluated various CNN architectures, such as AlexNet, VGGNet-16, and GoogleNet, for aerial scene classification tasks. They demonstrated that fine-tuning these pre-trained models on specific remote sensing datasets, like NWPU-RESISC45, significantly enhanced classification performance compared to using them as fixed feature extractors. This underscored the importance of task-specific fine-tuning when adapting CNNs to the unique characteristics of aerial imagery.

Qi et al. (2020) proposed MLRSNet, a multi-label dataset tailored to multi-scene aerial recognition, improving the evaluation of such tasks. The authors introduced MLRSNet, a large-scale dataset designed to address the limitations of single-label annotations in remote sensing scene classification.

Hua et al. (2019) used relation networks to model object interdependencies for more accurate multi-label predictions. They proposed a novel attention-aware label relational reasoning network designed for multi-label aerial image classification. The authors evaluated their model on the UCM multi-label dataset and a newly produced AID multi-label dataset, demonstrating its effectiveness.

Li et al. (2022) integrated spatial and channel-wise attention with CNN-LSTM for better context modeling. They proposed a novel deep learning framework that combined convolutional neural networks (CNNs) with long short-term memory (LSTM) networks, enhanced by spatial and channel-wise attention mechanisms. This architecture, termed the Residual Multi-Attentional CNN-LSTM Network, aimed to improve multi-label aerial scene classification by effectively modeling contextual dependencies and label correlations. The integration of attention mechanisms allows the model to focus on salient spatial regions and feature channels, thereby enhancing feature representation and classification performance.

Liu et al. (2022) introduced self-attention augmented CNNs, outperforming standard CNNs by capturing global dependencies. They introduced ConvNeXt, a pure ConvNet architecture that incorporated design elements inspired by vision transformers (ViTs), such as large kernel sizes, inverted bottlenecks, and layer normalization. While ConvNeXt did not explicitly integrate self-attention mechanisms, its design choices aimed to capture long-range dependencies and global context, traditionally associated with self-attention in transformers.

Tseng (2023) proposed a CNN-ViT ensemble, achieving robust performance across diverse aerial scenes. The author proposed a genetic algorithm (GA)-based weighted ensemble learning method that combined convolutional neural networks (CNNs) and vision transformers (ViTs) for multi-label aerial image classification. The ensemble integrated two CNN models and two ViT models, leveraging their complementary strengths to enhance classification performance. The GA optimized the weights assigned to each model in the ensemble, aiming to improve generalization and accuracy. Experiments conducted on the UCM and MLRSNet datasets demonstrated that the GA-based ensemble outperforms individual models and manually weighted ensembles, highlighting its robustness across diverse aerial scenes.

Mudraje et al. (2025) investigated the impact of image super-resolution (SR) techniques on enhancing multi-label scene classification in remote sensing applications. The authors explored the efficacy of SR as a pre-processing step to improve the quality of satellite images, thereby enhancing downstream classification performance. They evaluated four SR models—SRResNet, HAT, SeeSR, and RealESRGAN—and assess their impact on multi-label scene classification across various CNN architectures, including ResNet-50, ResNet-101, ResNet-152, and Inception-v4. The study found that applying SR significantly improves classification performance across various metrics, demonstrating its ability to preserve spatial details critical for multi-label tasks. This work offered valuable insights into selecting SR techniques for multi-label prediction in remote sensing and presented an easy-to-integrate framework to improve existing RS systems.

Li et al. (2020) introduced a hybrid framework that integrated CNNs with Graph Neural Networks (GNNs) to capture both visual features and spatial relationships among scene elements, enhancing multi-label classification performance.

MAIN FOCUS OF THE CHAPTER

Satellites capture aerial images from great distances. These images contain not only a single scene but also many objects in the landscape. This macroscopic view of single object classification may sometimes result in ignoring the major issue to be addressed by rendering the model infeasible in providing a comprehensive view of scenes in images for mission critical applications. This work focuses on the identification of all the scenes in a single aerial image using various CNN.

PROPOSED METHODOLOGY

Scene Protype Learning

Each aerial scene is divided into a fixed number of non-overlapping patches, with a patch size of 224x224. The number of patches is fixed for all scenes to 36 in this work.

A pre-trained DenseNet-121 CNN is used to extract features from each patch. Specifically, the DenseNet-121 model is fine-tuned on the training set of the dataset to extract features that are relevant for aerial scene recognition. The output of the ResNet-50 model is a feature vector of size 2048 for each patch.

After feature extraction, prototype representations are generated for each scene by averaging the features from all the patches. However, to capture the variability within each scene, the patch shuffling technique is used. Specifically, for each scene, a fixed number of prototypes (set to 10) by randomly shuffling the order of patches and averaging the features are generated.

To further improve the discriminative power of the prototype representations, a contrastive loss term is introduced during training. The contrastive loss encourages prototype representations of the same scene to be closer together in the feature space, while pushing prototype representations of different scenes further apart. The contrastive loss is defined as in Eq (1)

L = 1/2N * sum_i(sum_j(y_ij * d_ij^2 + (1-y_ij) * max(margin - d_ij, 0)^2))

(1)

where N is the total number of prototypes, y_ij is a binary variable indicating whether prototypes i and j belong to the same scene or not, d_ij is the Euclidean distance between prototypes i and j, and margin is a hyperparameter that controls the margin between prototypes of different scenes. The overall system design of the proposed prototype-based memory network is shown in figure 2. The prototype-Based Memory Network is a memory-augmented neural network that learns to store, retrieve, and compare information using class prototypes — representative feature vectors (centroids) for each class — in an external memory. These prototypes guide the classification of new, unseen inputs by comparing them to stored representations.

Figure 2. The overall system design of the prototype based memory network

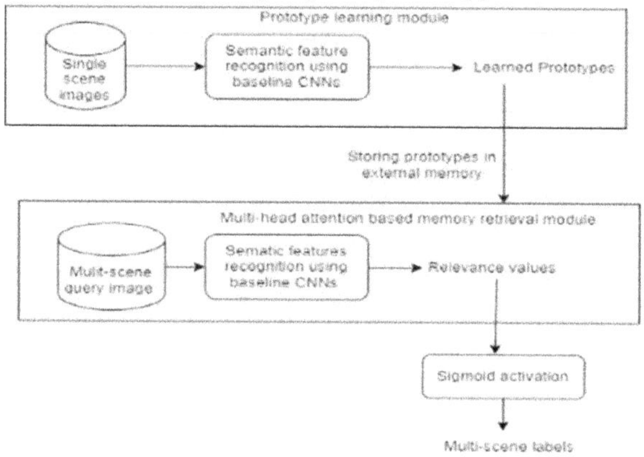

Figure 3. The network architecture of the prototype learning module

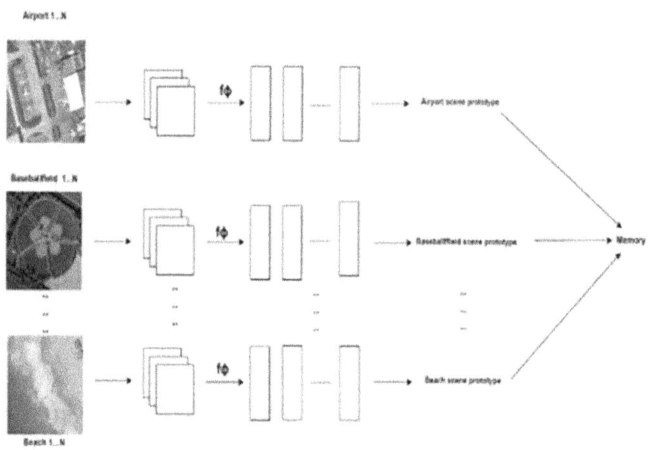

The network architecture of the prototype learning module is visualized in figure 3. In Figure 3, the prototype representations which are generated by the scene prototype learning module are stored in external memory. Specifically, the prototype representations are stored in a memory matrix M of size N x 2048, where N is the total number of prototypes.

Overall, the scene prototype learning module aims to generate informative and discriminative prototype representations of aerial scenes that capture the variability within each scene. These prototype representations are then stored in external memory.

Multi-Head Attention-Based Memory Retrieval

The first step in implementing the memory retrieval module is to prepare the data. Here, in this work, aerial images dataset was used and was divided into training and testing set. The images were pre-processed to extract features using a pre-trained Convolutional Neural Network (CNN) and then fed to the memory retrieval module. System design of proposed work is explained in figure 4. This figure 4 shows the procedure for the creation of multiple scene labels from the input multi-scene aerial image.

Figure 4. System design of proposed work

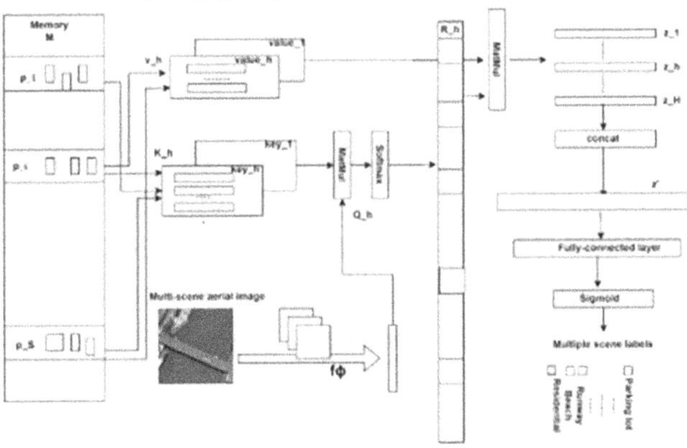

The next step is to generate prototypes for each class using a clustering algorithm. K-means clustering algorithm is used to generate prototypes, where the number of clusters is equal to the number of classes in the dataset.

The memory retrieval module uses a multi-head attention mechanism to retrieve relevant prototypes from memory. The input to the module is the features extracted from the pre-trained CNN. The attention mechanism computes a weight for each prototype based on its similarity to the input features. The weights are then used to compute a weighted sum of the prototypes, which serves as the output of the module.

The final step is to use the output of the memory retrieval module for classification. The softmax classifier assigns a class label to the input image based on the output of the module.

EXPERIMENTAL RESULTS

Dataset Description and Configuration

MAI Dataset

The Multiscene Aerial Image (MAI) dataset proposed by Hua et al. (2021) is a dataset for multi-scene aerial image analysis. The dataset consists of 10 scene categories, including urban, farmland, river, forest, and railway. This dataset was used for evaluating the proposed multi-scale spatial information fusion approach and adversarial learning technique for multi-scene classification. This dataset comprises 5,000 RGB images acquired by unmanned aerial vehicles (UAVs) with a spatial resolution of 0.1 meters per pixel. Each class has 500 images, and all images have a size of 1024x1024 pixels. The dataset offers predefined training, validation, and testing splits to facilitate the evaluation of multi-scene classification algorithms. Table 1. consists of the number of images associated with each scene. Figure 5 shows the sample images of MAI dataset. It shows sample images with multiple scene labels.

Figure 5. The example images from the MAI dataset (A) storage tanks, farmland (B) residential, farmland (C) runway, meadow (D) stadium, parking (E) beach, river (F) river, forest

UCM Dataset

The UCM dataset is a widely used dataset for remote sensing image analysis, particularly for land-use classification. The dataset consists of 21 classes of land-use categories, such as agriculture, forest, and residential areas. The images were acquired by the Landsat satellite system and have a spatial resolution of 30 meters per pixel. The dataset includes 2,100 images in total, with 100 images per class. The images are in RGB format and have a size of 256x256 pixels. The images were pre-processed to remove areas of clouds, shadows, and water bodies, resulting in clean land-use images. The sample images from UCM dataset are shown in figure 6. It shows sample images with unique scene label.

Figure 6. The example images from the UCM dataset (A) agricultural (B) buildings (C) chaparral (D) forest (E)golf course (F) overpass (G) tennis court (H) intersection

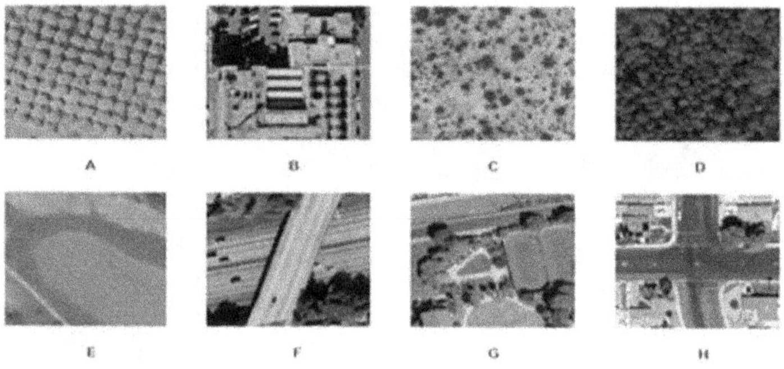

The dataset also provides a set of pre-defined training and testing splits, which have been widely used in the literature for evaluating the performance of land-use classification algorithms.

The UCM dataset has been used in various works for evaluating the performance of land-use classification algorithms, including deep learning-based approaches. The dataset's popularity is due to its large number of images, fine-grained land-use categories, and pre-defined splits that facilitate the comparison of different algorithms. 80% of the images from UCM dataset were selected for training purpose in order to learn the scene prototypes and remaining 20% images were used for testing purpose.

AID Dataset

Remote sensing image analysis, particularly for scene classification, often utilizes the Aerial Image Dataset (AID). This dataset includes 30 different scene classes, such as airport, harbor, and residential areas, and is commonly used in the field. The images were acquired by Google Earth and have a spatial resolution ranging from 0.5 to 2 meters per pixel. The dataset includes 10,000 images in total, with 200 to 800 images per class.

The images are in RGB format and have varying sizes, ranging from 600x600 to 4000x4000 pixels. The dataset also provides a set of pre-defined training and testing splits, which have been widely used in the literature for evaluating the performance of scene classification algorithms. The dataset's popularity is due to its large number of images, fine-grained scene categories, and varying image sizes, which pose a challenge for scene classification algorithms. Figure 7 shows the sample images of AID dataset. It shows sample images with unique scene label.

Figure 7. Sample images from AID dataset (A) airport (B) baseball field (C) beach (D) bridge (E) railway station (F) residential (G) viaduct (H) desert

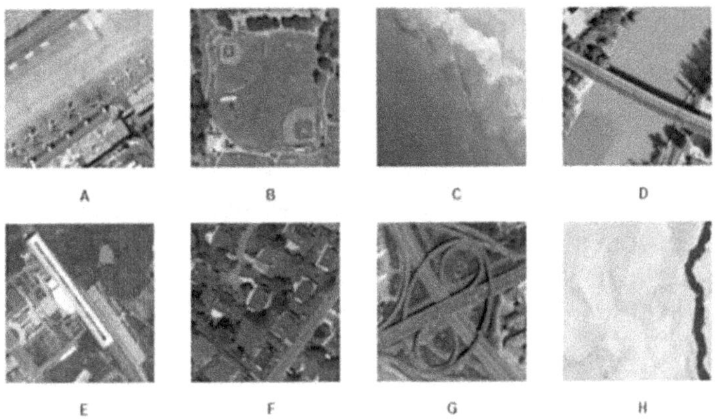

Implementation Details

To assess the performance of the proposed work, the embedding function is implemented with various deep CNNs.

Prototype Memory Network with VGGNet16

The prototype memory network is constructed using the VGGNet-16 baseline model by substituting all layers after the last max-pooling layer with a global average pooling layer. The size of the feat channel is set as 512.

Prototype Memory Network with ResNet50

The baseline deep CNN ResNet-50 is altered by removing layers after the global average pooling layer and the remaining layers are used as the embedding function. The feat channel size is set as 2048.

Prototype Memory Network with DenseNet121

The baseline deep CNN DenseNet-50 is altered by removing layers after the global average pooling layer and the remaining layers are used as the embedding function. The feat channel size is set as 2048.

The baseline CNNs are trained on the single-scene image datasets viz., UCM, AID and then they are taken as embedding functions. The obtained prototypes are concatenated to arrive at the memory M.

Training Details

The training process of the network involved utilizing the Adam optimizer with a learning rate of 0.0005 and a batch size of 32. To address the class imbalance issue in the dataset, a weighted cross-entropy loss function is adopted.

To prevent overfitting, the techniques such as data augmentation, random rotations, translations, and flips are applied to the training images. Also, early stopping is done based on the validation accuracy to prevent the model from overfitting to the training data.

During training, the authors used the UCM2MAI and AID2MAI datasets to train and validate the model, respectively. They also used the MAI dataset as a source of prototypes for the prototype learning module. The authors trained the model for 100 epochs and evaluated its performance on the AID2MAI dataset using both top-1 and top-5 accuracy metrics. They also compared their proposed model to several baseline models and demonstrated that their model outperformed the baselines on the AID2MAI dataset. Overall, the authors provided a comprehensive description of the training process for their proposed prototype-based memory network, including details on the optimizer, loss function, data augmentation, and evaluation metrics.

Table 1. Number of scenes per each category in UCM2MAI & AID2MAI datasets

Scene category	UCM2MAI		AID2MAI	
	UCM	MAI	AID	MAI
Apron	100	194	360	54
Baseball field	100	75	220	235
Beach	100	94	400	130
Commercial	100	607	350	1391
Farmland	100	680	370	983
Woodland	100	762	250	1312
Parking lot	100	708	390	1777
Port	100	3	380	9
Residential	200	958	700	2082
River	100	209	410	686
Storage tanks	100	89	360	193
Sea	100*	51	400*	59
Golf course	100	75	-	-
Runway	100	230	-	-
Sparse shrub	100	336	-	-
Tennis court	100	114	-	-
Bridge	-	-	360	878
Lake	-	-	420	756
Park	-	-	350	638
Roundabout	-	-	420	756
Soccer field	-	-	370	302
Stadium	-	-	290	136
Train station	-	-	260	9
Works	-	-	390	186
All	1600	1649	7050	3239

Performance Metrics

To quantify network performance, example-based F1- and F2-scores are used as evaluation metrics, which are calculated using Eq. 2

$$F_{\beta} = (1 + \beta^2)\frac{p_e r_e}{\beta^2 p_e + r_e}, \beta = 1,2 \quad (2)$$

where p_e and r_e denote precision and recall based on examples, they are calculated as in Eq. (3) and (4).

$$p_e = \frac{TP_e}{TP_e + FP_e} \quad (3)$$

$$r_e = \frac{TP_e}{TP_e + FN_E} \quad (4)$$

Experimental Results Across Models

UCM2MAI

To assess the performance of our models, we utilized two variations of the dataset and presented the qualitative outcomes in table 3. It can be seen that PM-DenseNet achieves the best mean F_1 and mean F_2 scores of 57.68 and 55.08 respectively. It also gains high performance and robustness in precision and recall with achieving scores of $m.P_e$ (70.04), $m.P_l$ (54.87) and $m.R_e$ (56.41), $m.R_l$ (29.93) which is better than its PM-VGGNet, PM-ResNet counterparts. Table 2 shows the example images with their predicted labels.

AID2MAI

To assess the performance of our models, we utilized two variations of the dataset and presented the qualitative outcomes in table 5. It can be seen that PM-DenseNet achieves the best mean F_1 and mean F_2 scores of 58.68 and 56.98 respectively. It also gains high performance and robustness in precision and recall with achieving scores of $m.P_e$ (71.24), $m.P_l$ (53.53).

While PM-ResNet achieves the highest of $m.R_e$ with the score of 55.34 and PM-DenseNet reaches upto 29.93 in $m.R_l$ score. Table 4 shows the sample images with their predicted labels.

Table 2. Sample images from UCM dataset and their predictions

Sample multi-scene aerial images from MAI dataset	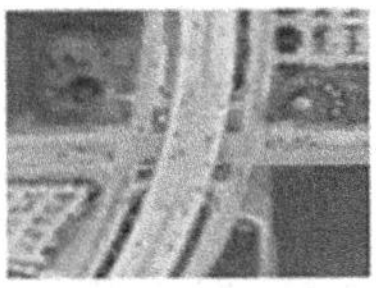		
Ground truths	Freeway, Parkinglot, Woodland	Storage tanks, River, Farmland	Beach, Residential, Woodland
Predictions	Freeway, Parkinglot, Woodland	Storage tanks, River, Farmland	Beach, Residential, Woodland
Sample multi-scene aerial images from MAI dataset	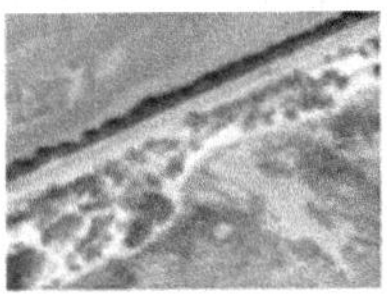	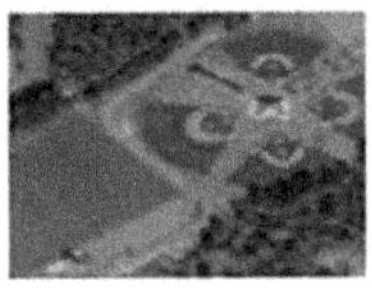	
Ground truths	River, Beach	Baseballfield, Woodland, Farmland	Stadium, Parkinglot
Predictions	River, Beach, Bridge	Baseballfield, Woodland, Farmland	Stadium, Parkinglot, Woodland

Table 3. Results on UCM2MAI dataset

Model	**m.F1**	**m.F2**	**m.Pe**	**m.PI**	**m.Re**	**m.RI**
PM-VGGNet	54.42	51.16	67.35	49.95	47.24	26.79
PM-ResNet	56.89	54.11	69.85	53.38	55.93	29.76
PM-DenseNet	57.68	55.08	70.04	54.87	56.41	29.93

Table 4. Sample images from AID dataset and their predictions

Sample multi-scene aerial images from MAI dataset	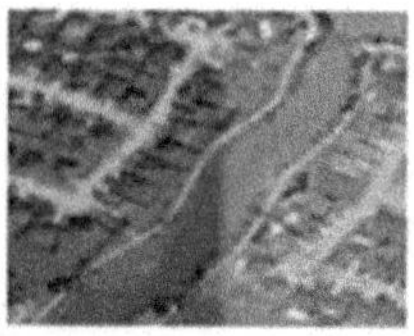		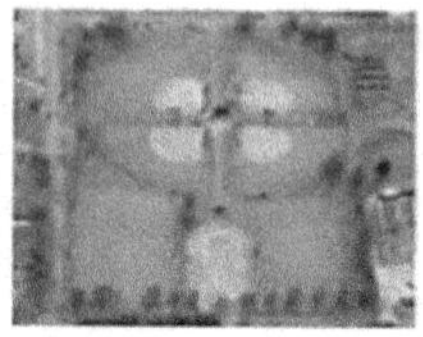
Ground truths	Bridge, Residential	Residential Woodland	Baseball, Parking lot
Predictions	Bridge, Residential	Residential Woodland	Baseball, Parking lot
Sample multi-scene aerial images from MAI dataset	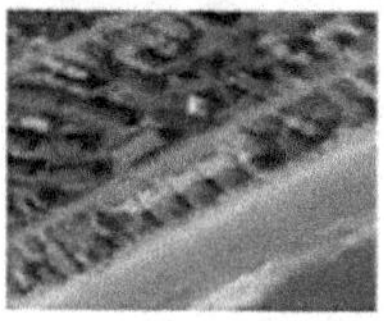		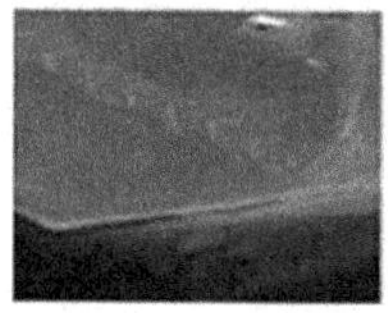
Ground truths	Residential, Beach	Bridge, River, Residential, Farmland	Runway
Predictions	Residential, Beach	Bridge, River, Residential, Farmland	Runway

Table 5. Results on AID2MAI dataset

Model	m.F1	m.F2	m.Pe	m.PI	m.Re	m.RI
PM-VGGNet	54.37	51.44	65.69	50.39	48.06	22.40
PM-ResNet	57.42	54.34	70.62	53.33	55.34	29.55
PM-DenseNet	58.68	56.98	71.24	53.53	54.21	29.93

FUTURE RESEARCH DIRECTIONS

This system has used the publicly available database for multiscene recognition. To make it more customized for particular application, the self-built or custom database can be created and tested.

CONCLUSION

To recognize multiple scenes from a single aerial image, three different datasets namely UCM dataset, AID dataset and MAI dataset consisting of 2100, 10000, 3923 images respectively were used. Out of the three datasets, UCM and AID dataset are Single scene datasets and the MAI dataset is a multi-scene dataset. These datasets are utilized in two configurations namely UCM2MAI and AID2MAI. Prototype based memory network using different CNN baseline models is used as the backbone. First, the single scene datasets were used to extensively train different baseline models and store them as prototypes in the external memory. Then a query image from the multi-scene dataset is given as input to the network to determine all the scenes present in it. Experiments were conducted to analyse the performance of different baseline models. From the experiments, it is concluded that PM-DenseNet outperforms the PM-VGGNet and PM-ResNet.

REFERENCES

Cheng, G., Han, J., & Lu, X. (2017). Remote Sensing Image Scene Classification: Benchmark and State of the Art. *Proceedings of the IEEE, 105*(10), 1865–1883. DOI: 10.1109/JPROC.2017.2675998

Eslami, S. M. A., Heess, N., Weber, T., Tassa, Y., Szepesvari, D., Kavukcuoglu, K., & Hinton, G. E. "Attend, Infer, Repeat: Fast Scene Understanding with Generative Models", *30th Conference on Neural Information Processing Systems (NIPS 2016)*, Barcelona, Spain, 2016.

Hariharan, B., & Ross, B. Girshick, "Low-Shot Visual Recognition by Shrinking and Hallucinating Features", *International Conference on Computer Vision (ICCV)*, pp. 3037-3046, 2017.

Hua, Y., Mou, L., & Zhu, X. X. (2019). Relation Network for Multi-label Aerial Image Classification. *arXiv preprint arXiv:1907.07274.*

Li, P., Chen, P., & Zhang, D. (2022). Cross-Modal Feature Representation Learning and Label Graph Mining in a Residual Multi-Attentional CNN-LSTM Network for Multi-Label Aerial Scene Classification. *Remote Sensing (Basel), 14*(10), 2424. DOI: 10.3390/rs14102424

Li, Y., Chen, R., Zhang, Y., Zhang, M., & Chen, L. (2020). Multi-label remote sensing image scene classification by combining a convolutional neural network and a graph neural network. *Remote Sensing (Basel), 12*(23), 4003. DOI: 10.3390/rs12234003

Liu, Z., Mao, H., Wu, C.-Y., Feichtenhofer, C., Darrell, T., & Xie, S. (2022). A ConvNet for the 2020s. *Proceedings of the IEEE/CVF Conference on Computer Vision and Pattern Recognition (CVPR)*, 11976–11986. https://doi.org/DOI: 10.1109/CVPR52688.2022.01170

Mudraje, A., Moser, B. B., Frolov, S., & Dengel, A. "Multi-label scene classification in remote sensing benefits from image super-resolution," *arXiv preprint arXiv:2501.06720*, Jan. 2025.

Qi, X., Zhu, P., Wang, Y., Zhang, L., Peng, J., Wu, M., Chen, J., Zhao, X., Zang, N., & Mathiopoulos, P. T. (2020). MLRSNet: A multi-label high spatial resolution remote sensing dataset for semantic scene understanding. *ISPRS Journal of Photogrammetry and Remote Sensing*, *169*, 337–350. DOI: 10.1016/j.isprsjprs.2020.09.020

Snell, J., Swersky, K., & Zemel, R. "Prototypical networks for few-shot learning" Advances in Neural Information Processing Systems (NeurIPS), 2017.

Sukhbaatar, S., Szlam, A., Welson, J., & Fergus, R. "End-To-End Memory Networks", arXiv:1503.08895, doi:/arXiv.1503.08895, 2015DOI: 10.48550

Sun, H., Li, S., Zheng, X., & Lu, X. (2020). Remote sensing scene classification by gated bidirectional network. *IEEE Transactions on Geoscience and Remote Sensing*, *58*(1), 82–96. DOI: 10.1109/TGRS.2019.2931801

Tseng, M.-H. (2023). GA-based weighted ensemble learning for multi-label aerial image classification using convolutional neural networks and vision transformers. *Machine Learning: Science and Technology*, *4*(4), 045045. DOI: 10.1088/2632-2153/ad10cf

Vinyals, O., Blundell, C., Lillicrap, T., Kavukcuoglu, K., & Wierstra, D. "Matching networks for one shot learning", *Proceedings of the 30th International Conference on Neural Information Processing Systems*, pp. 3637–3645, 2016.

Zhou, B., Lapedriza, A., Khosla, A., Oliva, A., & Torralba, A. (2018). Places: A 10 million Image Database for Scene Recognition. *IEEE Transactions on Pattern Analysis and Machine Intelligence*, *40*(6), 1452–1464. DOI: 10.1109/TPAMI.2017.2723009 PMID: 28692961

Chapter 6
Role of Machine Learning in Computational Fluid Dynamics

G. Prasad
https://orcid.org/0000-0002-5709-9182
Chandigarh University, Punjab, India

Snehal Malik
https://orcid.org/0009-0006-0823-1446
Chandigarh University, Punjab, India

Aadya Gupta
https://orcid.org/0009-0005-3149-6584
Chandigarh University, Punjab, India

Yash Nigam
https://orcid.org/0009-0006-5992-4023
Chandigarh University, Punjab, India

ABSTRACT

The integration of machine learning (ML) with computational fluid dynamics

DOI: 10.4018/979-8-3693-7525-9.ch006

(CFD) marks a significant advancement in the simulation and analysis of fluid flows. This chapter explores the synergistic role of machine learning in enhancing CFD methodologies, focusing on applications in modeling, optimization, and real-time analysis. Machine learning algorithms, particularly deep learning, offer powerful tools for identifying patterns and correlations within large datasets generated by CFD simulations. These algorithms can be trained to predict fluid behavior, accelerate simulation processes, and improve the accuracy of models by learning from empirical data. In modeling, ML techniques reduce the reliance on traditional empirical models, offering more precise and computationally efficient alternatives. Furthermore, ML-driven optimization techniques enhance the design process of fluid systems by enabling rapid evaluation of multiple design variables. Real-time data processing and analysis facilitated by ML also support adaptive control and decision-making in dynamic fluid environments.

INTRODUCTION

New solutions to tough problems which would have taken years and been expensive using traditional aircraft design techniques are being managed in weeks or months by means of machine learning. However, one of its most attractive potential use cases is for Computational Fluid Dynamics (CFD), a bedrock subsystem in understanding and optimising airplane designs. While traditional fluid simulations with machine learning can be used to boost the accuracy and speed of tasks which have a very different distribution from their training data (with hyper-parameter tuning, wavelet-like sensing for turbulence modelling, injection/acceleration emulation in fast-running shock physics simulators or natural convection as presented previously by improving thermal boundary layer design parameters). This chapter unveils the potential and ability of ML-powered large scale physical modelling tasks such as creating an airplane, forecasting climate reliefs. The same is true in expanded efforts, where these solutions are reaching beyond immediate predictive maintenance use cases to overflight congested areas of urban landscape inspection or remotely piloted flight controls and capabilities.

BACKGROUND OF THE RESEARCH

Recent advancements in artificial intelligence (AI) algorithms have significantly influenced the field of computational fluid dynamics (CFD), as highlighted by various researchers. B. Wang et al. (2021) examined the applications of AI across five key areas, including aerodynamic and turbulence models, as well as specific flow scenarios and mass and heat transfer processes. Barkalov, K. et al. (2022) focused on slope flow simulations, seeking optimal parameter values for their mathematical models. In parallel, Beetham, S. and colleagues (2020) developed a data-driven framework aimed at improving the closure formulations of the Reynolds-averaged Navier-Stokes (RANS) equations. Bounds, C.P. et al. (2023) discussed the integration of machine learning tools to enhance the prediction accuracy of CFD in external aerodynamic workflows. Similarly, Dangeti, C.S. (2018) emphasized the prevalence of RANS turbulence modeling in the automotive industry, while Frey Marioni, Y. (2021) and his team explored the use of Direct Numerical Simulation (DNS) data combined with machine learning to aid in RANS model development. Fu C. et. al (2017) identified RANS CFD simulations as critical first approximation tools for aerodynamics evaluations in motorsports. Further studies include Guilmineau, E. et al.'s investigation (2018) into RANS and Detached Eddy Simulation (DES) models for the Ahmed body and Eivazi H. et al.'s proposal of a deep neural network architecture for analyzing turbulent-flow data (2022). This growing body of literature underscores the collaborative potential of AI and traditional fluid dynamics methodologies in advancing the understanding and application of turbulence modelling. Eivazi H. et al. (2022) proposed a deep neural network architecture for learning a set of non-linear modes from high-fidelity turbulent-flow data useful for flow analysis.

H. G. et. al (1998) presented a comprehensive article detailing the principles of the FOAM C++ class library, which serves as a robust tool for continuum mechanics applications. This foundational work highlights the library's capabilities in facilitating complex simulations and enhancing computational fluid dynamics (CFD) research. Complementing this, Jeon J. and S. J. Kim (2022) emphasized the necessity of developing a neural network design specifically tailored for CFD, addressing the growing demand for more efficient and accurate predictive modeling in fluid dynamics. Furthermore, Ling J., A. Kurzawski, and J. Templeton (2016) contributed to the field by employing deep neural networks to average Reynolds turbulence modeling, integrating embedded invariance to improve the reliability and performance of turbulence predictions. Together, these studies reflect a significant advancement in the intersection of computational methods and machine learning within the realm of fluid dynamics. J. Wang his colleagues (2021) discussed how Reynolds-averaged Navier–Stokes equations with explicit data-driven Reynolds stress closure can be ill-conditioned.

Recent advancements in turbulence modeling and fluid dynamics have been significantly influenced by machine learning techniques, as evidenced by various studies. Liu W. et al. (2021) developed an iterative machine-learning framework specifically for Reynolds-Averaged Navier-Stokes (RANS) turbulence modeling, while Brenner et al. (2019) provided a comprehensive perspective on the role of machine learning in advancing fluid mechanics. Obiols-Sales et al. (2020) introduced "CFDNet," a deep learning-based accelerator aimed at enhancing fluid simulations. Pope S. B.'s work on turbulence modelling and Spalart's strategies (1999) further contributed to understanding these complex phenomena, including the development of a one-equation turbulence model for aerodynamic flows by Spalart and Allmaras (1992). Historical perspectives also play a crucial role; Kraichnan's 1967 paper on two-dimensional turbulence established the concept of dual cascades, which remains fundamental in the field. More recent contributions, such as Maulik et al.'s exploration of neural networks for subgrid modelling (2019) and Vinuesa R. et al.'s investigation (2022) of pressure-gradient turbulent boundary layers, continue to enrich our understanding of turbulent flows. Additionally, Godunov and Bohachevsky's pioneering (1959) finite difference method laid the groundwork for modern computational fluid dynamics, while Shi et al. highlighted the synergy between CFD, machine learning, and experimental analysis for optimizing thermal plasma reactors. Brunton S. L. and Kutz J. N.'s book (2022) offers a broad overview of data-driven science and engineering, and Poroseva S. V. et al. (2016) assessed RANS simulation accuracy against direct numerical simulation (DNS) data, identifying key areas for improvement. Lastly, Shan et al.'s 2020 paper proposed a deep learning-based fast solver for Poisson's equation, underscoring the transformative potential of AI in accelerating numerical simulations across diverse applications.

FUNDAMENTALS OF MACHINE LEARNING IN CFD

1. **Supervised learning**: In supervised learning, a dataset is trained to an algorithm in which the chosen features of a particular case are usually used as the desired result; these can be the lift coefficient and the drag coefficient with the input signals such as temperature, pressure, and velocity of the flow are required. The purpose of this is to create a function that is learned based on the training data that can predict the result of new instances never seen before.
2. **Un-supervised learning**: Basically, on this topic, all the known algorithms don't have any data or any label data on which they train the proven. They will figure it out themselves based on the information given to them to distinguish the mediators, structure or abuse in the data.

TYPES OF MACHINE LEARNING

Which algorithm to choose depends on the CFD job, data type and accuracy required. The different CAE tools become more valuable under the hands of a designer and engineers than they are in producing numerical outputs. This is one of the main reasons why combat fighters can assimilate advanced data from the CFD method to recommend optimization of the design and the whole system when machine learning algorithms are run, but only if they are good enough, they can be put into practice.

1) The regression algorithms that predict the continuous numerical values are regression algorithms, in general. The regression algorithms can be used in CFD for:
   a. Predicting flow properties: For instance, determining pressure, velocity, temperature at specific points or positions

b. Optimal design parameters: Optimal design parameters are those which are the best values that the elements belonging to that design take to give the desired performance measure. Some of the more widely known algorithms used in regression include:
    - Linear regression: The linear correlation between the two variables (input & output) is assumed.
    - Support Vector Regression (SVR): A support vector machine is used to develop a linear or nonlinear regression model.
    - Random Forest Regression: An ensemble approach where multiple decision trees are usually trained to improve accuracy.
    - Neural Networks: It represents complex nonlinear relationships between input variables and output variables.

2) Algorithms for Classification: These are based on prediction with respect to categorical labels. In CFD, they can:
   a. Identify flow regimes: classify flow patterns as either laminar, turbulent, or transitional.
   b. Fault detection: recognize anomalies or defects in CFD simulations. The most widely used algorithms for classification are:
    - Logistic Regression: For two-class problems, for example, laminar or turbulent.
    - SVM: The algorithm definitely supports linear, as well as non-linear, classification.
    - Decision Trees: This will develop a tree for prediction by creating a sequence of yes or no questions.
    - Random Forests: It is an ensemble method that integrates several decision trees using improved accuracy.
    - Neural Networks: Applicable for regression and classification.

3) Clustering Algorithms: These algorithms are generally employed to cluster similar data points. In the scope of CFD, they can:
   a. Give information about the flow structure: group similar flow regions to form clusters based on certain properties.

b. Identify anomalies: expose flow types that are abnormal compared with the general flow features that might be indicative of a problem or defect. Some of the popular clustering algorithms are:
    - K-means clustering: A fast and simple spatial algorithm used for partitioning data into k number of clusters.
    - Hierarchical Clustering: Clusters of all sizes, from small to large.
    - DBSCAN-cluster: Based on density and noise.

DATA PRE-PROCESSING AND FEATURE ENGINEERING FOR CFD DATA

Data pre-processing and feature engineering are among the most fundamental steps carried out on the CFD data to improve dataset quality for machine learning purposes. Data pre-processing includes the cleaning of CFD data, removal of noise, outlier elimination, and filling of missing values, all starting from the need for the dataset to be reliable. The need for normalization as well as scaling is thus made to ensure that the data is put within a comparable range in order to train the model more successfully. Transformation techniques have used to smooth the data distribution and minimize complexity. Examples include log transformations and others belonging to the dimensionality reduction techniques such as PCA. Feature engineering will be the next step, with the objective of extracting and producing informative features that represent the underlying principles of fluid dynamics. This may involve computing derived quantities giving a more detailed insight into the behaviour of flows, including turbulence metrics, pressure gradients, and vorticity, though also, time-series features may be produced for the purpose of simulating temporal dynamics, and statistical summaries are capable of capturing many key aspects of the flow field. In this way, incorporating interaction terms and encoding categorical variables in the feature set facilitates learning for machine learning algorithms. Collectively, these steps assure well-prepared and improved CFD data, enhancing an overall performance of the model and predictability accuracy in applications involving fluid dynamics.

APPLICATIONS OF ML IN CFD

Recently, machine learning (ML) has attracted considerable attention for advancing the capabilities in Computational Fluid Dynamics (CFD) from many different perspectives such as forward modelling, inverse design & control and data analysis & visualization.

- **Forward Modelling**: To speed up CFD simulation using machine learning (ML) based approaches such reduced order/surrogate modelling for rapid prediction without incurring the high computational cost behind detailed simulations. This method allows engineers to explore their design spaces more efficiently.
- **Predictive Maintenance**: ML models are able to monitor data produced by CFD over time and detect flow-associated anomalies that could be signs of wear, blockage, or future failure in fluid systems (e.g., HVAC, turbines, or pipelines). Classification algorithms within the models are utilized to identify irregular patterns in the flow, time-series models for predicting failure, and clustering for detecting deviations from normal operational conditions.
- **In inverse design and control**: CFD simulation optimization flow control, form optimization (Machine learning methods) The above strategies increase the efficiency and efficacy in fluid dynamics applications by changing flow characteristics on-the-fly and predicting performance of various geometrical configurations.
- **Surrogate Modelling**: These models will help predict the outcome of entire CFD simulation for new input parameters after being trained on a small subset of CFD simulations. This will certainly accelerate the design and optimization process quite a bit.
- **Mesh Optimization and Adaptive Meshing**: Machine learning aids in refining mesh quality dynamically by learning from simulation data: adaptive meshing algorithms, mesh quality prediction, mesh-independent flow feature detection.
- **Turbulence Modelling**: Machine learning provides a data-driven strategy to model turbulence more precisely than conventional RANS models. Based on high-fidelity DNS or LES data, ML algorithms can forecast Reynolds stress components and intricate

flow interactions challenging for empirical models to represent. Prominent techniques are: Neural networks for Reynolds stress closure, sparse regression for turbulence modelling, hybrid RANS-ML frameworks for improved accuracy.

- **More detailed flows**: Machine learning (ML) could be utilized in order to generate data-driven turbulence models better performing and providing higher fidelity than traditional type of model. These models may learn complex interactions between flow variables when they are training on big datasets of CFD simulations or experimental data.

Figure 1. Data-driven model for improving RANS prediction

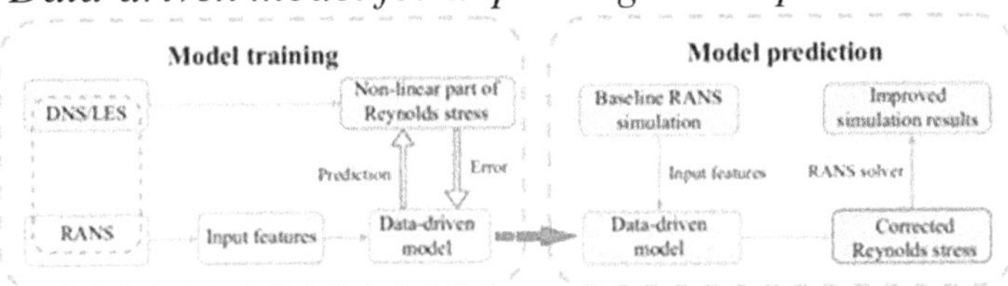

Figure 1 shows the possible use of the data-driven models and the classical Reynolds-Averaged Navier-Stokes (RANS) approaches towards the problem of turbulence flow prediction. This procedure improves the RANS simulations accuracy through training the model with high fidelity DNS/LES data. Further, the data-driven model wishes to bring improvement to RANS model predictions by concentrating on turbulent stresses and therefore bridge the gap with regard to accuracy between RANS and GALES which are fast but rather expensive respectively.

Right Side Model Prediction

- Baseline RANS Simulation: This problem of interest is solved using a standard RANS approach analysis.
- Input characteristics: The processed data driven model outputs its input characteristics acquired from the RANS EHD simulation.

- Data-Driven Model: These enhanced responses of the adjusting the Reynolds stresses were predicted by the data driven model rather than the baseline RANS data.

Training Model on Left Side

- Direct Numerical Simulation/Large Eddy Simulation or DNS/LES employs the use of numerical simulations which enable capturing small turbulent scales and is referred to as a high accuracy simulation method. It is accurate and so detailed.
- Reynolds-Averaged Navier-Stokes or RANS models do not perform better than DNS/LES models however, they are very much able to reproduce the turbulent flow field through the use of averaging.
- Features of Input: In order to train a data-driven model, features are extracted from both DNS/LES data and RANS data and loaded into the model as a training input. Non-linear Reynolds stress component: It is worth mentioning that using data-driven models, the nonlinear parts of the turbulent stresses that are very difficult for RANS to predict correctly can be handled.
- Data-driven Model: A machine learning model or other types of models in the Data-Driven approach can be trained and used to predict the nonlinear components of the Reynolds stresses. The difference between what was anticipated from the results of the simulations, and from DNS/LES results.
- Mesh creation: By using machine learning to automate and optimize the mesh creation process, simulation quality and efficiency may be increased while computing costs are decreased.

Figure 2. Adaptive Mesh Refinement (AMR)

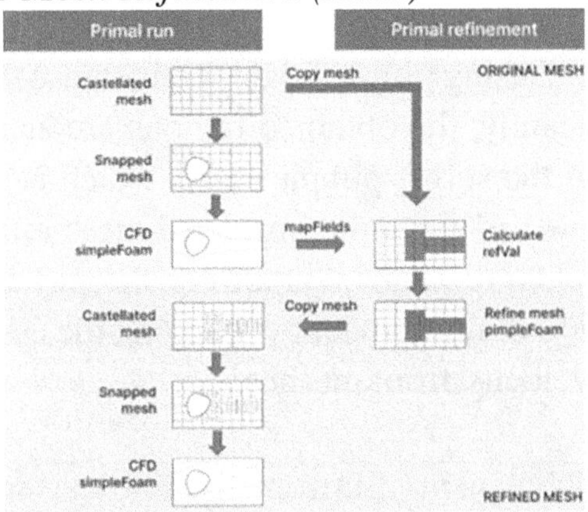

(Sadrehaghighi, 2022)

Figure 2 shows the workflow of CFD in the context of a meshing process, first in the generation and refinement phases of a primal run, followed by a further refinement. The graph described above shows a step-by-step, two-stage approach to improve the quality and precision of a simulation.

a. The first of these is called Castellated Mesh, and is the piece with sharp edges, representing one of the primary shapes that a meshing algorithm will produce.
b. In the second step, the mesh is snapped to a closer fit against the geometry surface; during this stage, the castellated mesh is "snapped" onto the geometric shape.
c. From this step onwards, the process of CFD Simulation using simpleFoam follows; the snapped mesh is then applied in the solver of CFD, namely simpleFoam, that it has been created for steady-state simulation of incompressible flow. It is noteworthy as at this step, the first set of results from the simulation process emerges.
d. Primal refinement process starts with the creation of the technique for refinement that duplicates the original mesh.
e. Then, there is a computation, refVal, to find where the more refinement is required. The computation may also be done along with the error estimation or any other relevant criteria for refinement. However,

note that whatever step is made here relies totally on how accurate and precise that is.

f. Mesh is improved using pimpleFoam. To increase the precision of the solution significantly, the obtained regions are again further refined with the help of the solver pimpleFoam, which is specially intended for transient flow and gives the improved mesh afterward.
g. Further, mapFields is used to maintain consistency and continuity with the whole process of refinement by mapping the essential fields, such as pressure or velocity from meshes.

Overall, this enrichment in CFD from machine learning not only streamlines the simulations but also facilitates much better design processes and improvement of data interpretation. Generally, this development opens up increasingly innovative solutions for engineering, but challenges abound.

COMPARATIVE ANALYSIS: TRADITIONAL CFD vs ML-AUGMENTED CFD

Machine Learning augments CFD by offering quick, scalable, and adaptive solutions, most useful in design optimization, uncertainty quantification, and exploratory analysis. But it brings new challenges due to data dependence, generalization, and physical interpretability that need to be weighed against the advantages of established methods as shown in Table 1.

Table 1. Comparison of ML and CFD

Aspect	Traditional CFD	ML-Augmented CFD
Computational Cost	High, especially for DNS or LES; simulations can take days	Significantly reduced through surrogate modelling or reduced-order models
Physical Consistency	Explicitly governed by Navier-Stokes and turbulence models	Can be less interpretable or physically inconsistent unless physics-informed
Scalability	Limited by numerical complexity and solver stability	Scales better with data availability; enables real-time inference

continued on following page

Table 1. Continued

Aspect	Traditional CFD	ML-Augmented CFD
Interpretability	Fully traceable due to reliance on known physics	Often treated as black-box unless explainable ML techniques used
Data Requirement	Minimal beyond boundary conditions and mesh	Requires large, high-fidelity datasets for training
Adaptability	Needs manual model tuning for new problems	Automatically adapts via retraining or transfer learning
Applications	Reliable for mature fields with established models	Expands feasibility in complex, nonlinear, or data-rich systems

STRATEGIES FOR SCALING ML- DRIVEN CFD TO INDUSTRY ADOPTION

In order to transition machine learning from the research laboratory to general industrial CFD processes, there are multiple strategic enablers that need to be solved. These methods enable scalability, trust, and operational integration within industries such as aerospace, automotive, energy, and manufacturing.

1. Standardized Benchmark Datasets: Developing high-scale, open-access CFD datasets enables developers and researchers to train, compare, and verify ML models in a reliable fashion. Analogous to ImageNet in computer vision, curated CFD benchmarks have the potential to drive innovation faster and provide consistency across applications.
2. Transfer Learning and Model Generalization: Industrial CFD problems differ in complexity, geometry, and boundary conditions. Transfer learning and domain adaptation methods assist in applying ML models learned from one setting to new environments, making them more reusable and adaptable.
3. Physics-Informed Machine Learning: Incorporating physical constraints (e.g., mass and momentum conservation) within learning algorithms guarantees that predictions are realistic and interpretable—essential in safety-critical domains such as aerospace. It also enhances the acceptability of ML models within regulatory frameworks.

4. Workflow Integration: ML-CFD software must seamlessly integrate with conventional CFD solvers, CAD tools, and optimization frameworks. APIs, plug-ins, and ML-augmented solvers must enable user-specified flexibility and inter-operability.
5. Cloud and Edge Computing for Scalability: Cloud-based deployment facilitates high-scale batch processing and team-wide collaboration. On the other hand, edge deployment (e.g., in UAVs or real-time monitoring applications) facilitates light-weight, fast inference for adaptive systems.
6. Industry-Academic Collaboration and Training: Workforce upskilling is imperative. Training across disciplines in CFD, AI, and data science, as well as collaboration between industry and academia, can equip engineers with the capability to implement ML-facilitated design and simulation tools at scale.

These initiatives are not just designed to enhance model performance but to make ML-CFD feasible, trustworthy, and scalable for actual engineering problems.

AIRCRAFT SPECIFIC APPLICATIONS OF ML IN CFD

By shortening simulation runtimes and facilitating quick design iterations, machine learning greatly shortens project timeframes in CFD processes (Glaessgen & Stargel (2012)). For example, early-stage idea validation cycles can be shortened from months to weeks by using surrogate models to efficiently explore aerodynamic design areas. By reducing the need for high-performance computers and accelerating physical prototyping, this efficiency directly results in cost reductions.

These benefits do, however, come with difficulties in terms of regulatory compliance. Transparency, traceability, and rigorous evaluation of simulation tools are required by certification agencies such as the FAA and EASA. Many machine learning models are "black-box" in nature, which makes verification and validation (V&V) more difficult and may cause certification attempts to be delayed. Furthermore, model reliability is a challenge because it is frequently difficult to collect high-quality training data for safety-critical systems. Integrating ML-enhanced CFD into certified design workflows—particularly for piloted aircraft—remains a significant challenge in the absence of thorough documentation and explainability. In order to overcome these obstacles in workflows of the future, explainable AI and machine learning informed by physics must be implemented.

The broad use of ML-CFD technologies in aircraft depends on regulatory approval in addition to performance and accuracy. Many machine learning models are black-boxed, which makes them incompatible with the traceability and explainability standards set by aircraft regulators such as the FAA and EASA. Comprehensive V&V is required by certification standards, however most neural network-based models currently struggle to meet these requirements. New frameworks built on physics-informed machine learning, strict documentation procedures, and the creation of interpretable models are all necessary to close this gap and guarantee confidence in high-stakes flying situations.

LIMITATIONS OF MACHINE LEARNINF IN CFD

1. Physical Consistency: Developing techniques that guarantee the ML models will conserve the basic equations of fluid motion is never a walk in the park. Non-physical characteristics or other inconsistencies can yield non-physical predictions.
2. The Quality and Quantity of Data: Constraints present in the data may place a limitation regarding the performances of the available ML models. Bad data, especially noisy data, may cause model inaccuracy and possess low model fidelity.
3. Interpretation and Explanatory Power: It is also hard to comprehend how certain decisions are made by more sophisticated models of ML. Transparency is a key aspect in acceptance of model predictions and contributes to the proper use of the models in engineering practice.
4. Compatibility: The problems may include those alluded to above like data and software incompatibilities, and inter-computing unit connectivity. Building ML models may be processor intensive and may take long time especially when the data employed is large or the designs

are complicated, even if ML is able to reduce the turnaround time for some applications.

5. Complexity of Fluid Flows: One of the problems is that turbulent flow, multiphase flow, and transient phenomena in fluid mechanics exert an enormous influence on the success of an ML application.

For many CFD applications, high-fidelity training data are difficult and expensive to generate. Such a CFD model can generate huge datasets; however, not all the possible outcomes may be included in these models. When the flows become extremely complex, then not all possible outcomes may be accounted for, and that data may not be available and thus may limit the effectiveness of ML models. The aforementioned limitations have a significant impact on enterprise-level adoption and are not just theoretical concerns. For example, in safety-focused fields like aerospace, where a single incorrect prediction can have catastrophic consequences, the lack of physical resilience in ML models can cause engineer skepticism. The use of ML-augmented tools in structured design cycles might also be hindered by uninterpretable or dark model behaviour, which can make regulatory certification more difficult.

Additionally, sectors often lack access to the huge, high-fidelity datasets needed to train models, particularly for novel or specialized designs, making the data consumption of ML models a scalability problem. This can hinder democratization by restricting the use of ML to big businesses with established simulation infrastructures. Finally, enterprise adoption is hampered by compatibility with outdated technologies and workflows, where cycles of digital transformation are sluggish and extremely risk averse. The use of ML in industrial CFD will be limited to experimental or non-critical applications unless these issues are resolved with targeted research and tools. To close this adoption gap, a comprehensive approach that combines explainability frameworks, modular toolchains, physics-informed machine learning, and industry standards is required.

FUTURE DIRECTIONS AND RESEARCH OPPORTUNITIES IN MACHINE LEARNING FOR CFD

Figure 3. Machine learning in CFD

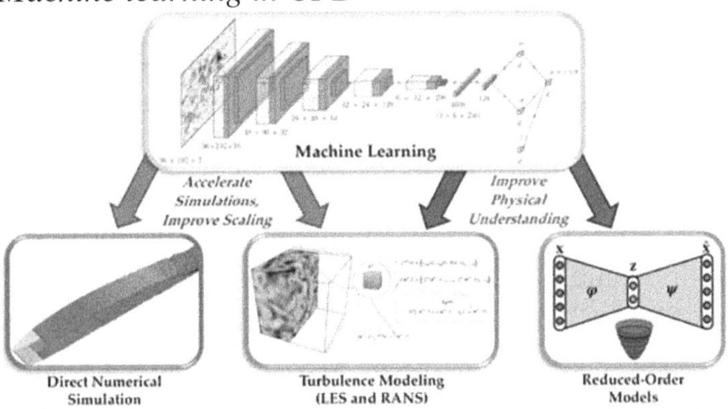

Figure 3 depicts Computational Fluid Dynamics (CFD) as the branch of technology that deals with the use of Machine Learning (ML) to enhance both the simulation techniques and physical understanding. In sum, the image seems to suggest that the application of ML will enhance the science of CFD through improving physical interpretation, enhancing simulation performance, and the rapid development of accurate and efficient computational models.

Upper Section

1. Mathematical Equations: These are arguably the Navier-Stokes equations among other governing equations of fluid dynamics.
2. Numerical Grid: The structure of a grid indicates the discretization of the computing domain for numerical simulations.
3. Convolutional Neural Network (CNN): This architecture has been designed to tackle problems like flow data feature extraction and picture identification using machine learning.

Bottom Section

1. Direct Numerical Simulation (DNS): These are very expensive computational simulations of turbulence which are able to resolve all relevant turbulent scales down to the Kolmogorov scale.
2. Turbulence Modelling (LES and RANS): These are approximation methods developed in such a way that they treat turbulence at different levels of detail in order to reduce computational expenses.
3. Reduced-Order Models: These are simplified versions of a system that typically contain all of the important dynamics but are developed with or without the aid of ML.
    1. **Real-Time CFD Simulations Using ML:** Current advancements have been geared toward using ML to speed up CFD solvers toward real-time or near-real-time predictions, especially in control and interactive design applications. Surrogate models, neural operators, and PINNs (Karniadakis et al. (2021)) are capable of achieving substantial speedup without loss of fidelity and are therefore effective in real-time flow control, adaptive design, and digital twin systems.
    2. **Multi-scale modelling**: Here, an extremely optimistic role will be offered by machine learning, enabling a seamless integration of scale disparity in CFD simulations. This technique, which unites macroscopic models with microscopic phenomena, can assist in providing an understanding of the complex fluid behaviour and the establishment of reliable forecasts across several applications.
    3. **Physics-based researches and machine learning**: Deployment of physical laws and restrictions to augment the machine learning model provides a reasonable corridor to enhance the precision and dependability of the CFD simulations. The physics-informed machine learning is able to furnish predictions upon the training of its model with some prior knowledge from fluid mechanics, such that they accord with basic principles and lend themselves to easy interpretability.
    4. **Real-Life Applications**: One of the takeaways from this unpredictability is to broaden the scope of the intended study to include more of potential uses of machine-learning applications in practical

CFD problems. Areas such as specialized engineering can be included in this where one can use the benefits of machine learning to gain insight on the flow of fluids within organisms mapping and biological systems. In addition, weather forecasting is one more area where augmented machine learning processes further the capability of forecasting by means of big data.

5. **Hybrid models- Combining Traditional CFD with ML:** One of the key frontiers in CFD research is hybrid framework development combining the strength of traditional solvers (e.g., RANS or LES) with the agility of ML. Such models promote prediction quality, lower computational cost, and provide physical consistency. Some examples include neural network closures for turbulence, ML-corrected wall functions, and data assimilation techniques where ML incrementally enhances classical solvers by online feedback
6. **Accelerating CFD Simulations**: Future work may include designing machine learning algorithms that would be used to facilitate real-time CFD simulations which could allow an engineer to predict fluid behaviour in real-time, thus enabling quick redesigning and prototyping Frame.
7. **ML for Sustainable Aviation and Autonomy**: Machine learning will drive innovation in aerospace, particularly towards sustainable aviation and autonomous flight systems. In the drive towards greener skies, ML-assisted CFD models allow for quick optimization of aerodynamic geometries for reducing drag and fuel consumption. Such tools also aid the creation of future-generation propulsion systems (such as electric and hybrid engines) by optimizing thermal management and flow distribution. In autonomous flight, ML-augmented CFD is helping create reliable flight control systems that provide accurate real-time prediction of aerodynamic loads, flow separation, and unsteady phenomena. Such knowledge is paramount in unmanned aerial vehicles (UAVs), urban air mobility (UAM) vehicles, and future autonomous commercial flight, where adaptive decision-making and fluid-structure interaction modelling are imperative (Tao et al. (2023)).

8. **Reinforcement Learning (RL) for Flow Control:** Reinforcement learning in CFD is increasingly being used as an approach to optimal control methods (Rabault et al. (2019)). By learning how to act on dynamic fluid situations (e.g., delay of separation, boundary layer management), reinforcement learning (RL) provides an innovative approach to optimizing control actions in real-time. This proves useful in active flow management in propulsion systems or wing applications.
9. **AI-Augmented Design Pipelines:** In addition to supervised learning, independent design space exploration is researched using AI methods such as generative models and neural architecture search. Such systems can continuously improve materials, boundary conditions, and geometries by leveraging a feedback loop between machine learning predictions and physical simulations.
10. **Digital Twins for CFD and Aero Systems:** Digital twins—virtual copies of physical systems combine CFD, sensor data in real time, and ML to allow for predictive monitoring and design feedback (Zhang et al. (2021)). For aero systems, digital twins of engines, airframes, or UAVs may employ ML-augmented CFD to model performance, forecast maintenance requirements, and modify operations in real time (G. Genevaux et al. (2022)).

CONCLUSION

There is an increasing adoption of machine learning (ML) techniques in computational fluid dynamics (CFD) which enhance the quality, efficiency, and understanding of the processes. The researchers and the engineers can implement some of the machine learning methods to accelerate the process of CFD in conducting the simulation. Machine learning can reduce the computational burden by conducting CFD computations through model order reduction and surrogate modelling; enhance designs using flow control/shape optimization; and mine large datasets for useful information. However, there are still many challenges that lie ahead even with the potential, such as preservation of physical consistency, data quality control, and improving qualitative aspects of the model.

It appears reasonable to anticipate that a completely fresh wave of advancements will be made in the application of machine learning to effective computational fluid dynamics. It will also be possible to make more comprehensive evaluations due to advances in strategies of multi-scale modelling that combine phenomena happening on multiple scales. The reliability of the models will improve because physics guiding machines will guarantee models are within the most fundamental physical principles. Similar to how it's believed that applying ML to search for new solutions to practical problems like medicine and climate change will also provide radical breakthroughs and solutions. It is also believed that these technologies will have an increasing bearing on CFD in the future that there will be many more engineering process optimizations which were previously thought impossible.

REFERENCES

Barkalov, K., Lebedev, I., Usova, M., Romanova, D., Ryazanov, D., & Strijhak, S. (2022). Optimization of turbulence model parameters using the global search method combined with machine learning. *Mathematics*, *10*(15), 2708. DOI: 10.3390/math10152708

Beetham, S., & Capecelatro, J. (2020). Formulating turbulence closures using sparse regression with embedded form invariance. *Physical Review Fluids*, *5*(8), 084611. DOI: 10.1103/PhysRevFluids.5.084611

Bounds, C. P., Uddin, M., & Desai, S. (2023). Tuning of Turbulence Model Closure Coefficients Using an Explainability Based Machine Learning Algorithm (No. 2023-01-0562). SAE Technical Paper.

Brenner, M. P., Eldredge, J. D., & Freund, J. B. (2019). Perspective on machine learning for advancing fluid mechanics. *Physical Review Fluids*, *4*(10), 100501. DOI: 10.1103/PhysRevFluids.4.100501

Brunton, S. L., & Kutz, J. N. (2022). *Data-driven science and engineering: Machine learning, dynamical systems, and control*. Cambridge University Press. DOI: 10.1017/9781009089517

Dangeti, C. S. (2018). Sensitivity of Turbulence Closure Coefficients on the Aerodynamic Predictions of Flow over a Simplified Road Vehicle (Master's thesis, The University of North Carolina at Charlotte).

Eivazi, H., Le Clainche, S., Hoyas, S., & Vinuesa, R. (2022). Towards extraction of orthogonal and parsimonious non-linear modes from turbulent flows. *Expert Systems with Applications*, *202*, 117038. DOI: 10.1016/j.eswa.2022.117038

Frey Marioni, Y., de Toledo Ortiz, E. A., Cassinelli, A., Montomoli, F., Adami, P., & Vazquez, R. (2021). A machine learning approach to improve turbulence modelling from DNS data using neural networks. International Journal of Turbomachinery. *Propulsion and Power*, *6*(2), 17.

Fu, C., Uddin, M., Robinson, C., Guzman, A., & Bailey, D. (2017). Turbulence models and model closure coefficients sensitivity of NASCAR Racecar RANS CFD aerodynamic predictions. SAE International Journal of Passenger Cars-Mechanical Systems, 10(2017-01-1547), 330-344.

Genevaux, G., Daniel, L., & Chevalier, T. (2022). A digital twin approach for predictive maintenance of aircraft systems.

Glaessgen, E., & Stargel, D. (2012). The digital twin paradigm for future NASA and US Air Force vehicles.

Godunov, S. K., & Bohachevsky, I. (1959). Finite difference method for numerical computation of discontinuous solutions of the equations of fluid dynamics. Matematičeskij sbornik, 47(3), 271-306.

Guilmineau, E., Deng, G. B., Leroyer, A., Queutey, P., Visonneau, M., & Wackers, J. (2018). Assessment of hybrid RANS-LES formulations for flow simulation around the Ahmed body. *Computers & Fluids*, *176*, 302–319. DOI: 10.1016/j.compfluid.2017.01.005

Jeon, J., Lee, J., & Kim, S. J. (2022). Finite volume method network for the acceleration of unsteady computational fluid dynamics: Non-reacting and reacting flows. *International Journal of Energy Research*, *46*(8), 10770–10795. DOI: 10.1002/er.7879

Karniadakis, G. E., Kevrekidis, I. G., Lu, L., Perdikaris, P., Wang, S., & Yang, L. (2021). Physics-informed machine learning. *Nature Reviews. Physics*, *3*(6), 422–440. DOI: 10.1038/s42254-021-00314-5

Kraichnan, R. H. (1967). Inertial ranges in two-dimensional turbulence. *The Physics of Fluids*, *10*(7), 1417–1423. DOI: 10.1063/1.1762301

Ling, J., Kurzawski, A., & Templeton, J. (2016). Reynolds averaged turbulence modelling using deep neural networks with embedded invariance. *Journal of Fluid Mechanics*, *807*, 155–166. DOI: 10.1017/jfm.2016.615

Liu, W., Fang, J., Rolfo, S., Moulinec, C., & Emerson, D. R. (2021). An iterative machine-learning framework for RANS turbulence modelling. *International Journal of Heat and Fluid Flow*, *90*, 108822. DOI: 10.1016/j.ijheatfluidflow.2021.108822

Maulik, R., San, O., Rasheed, A., & Vedula, P. (2019). Subgrid modelling for two-dimensional turbulence using neural networks. *Journal of Fluid Mechanics, 858*, 122–144. DOI: 10.1017/jfm.2018.770

Morita, Y., Rezaeiravesh, S., Tabatabaei, N., Vinuesa, R., Fukagata, K., & Schlatter, P. (2022). Applying Bayesian optimization with Gaussian process regression to computational fluid dynamics problems. *Journal of Computational Physics, 449*, 110788. DOI: 10.1016/j.jcp.2021.110788

Obiols-Sales, O., Vishnu, A., Malaya, N., & Chandramowliswharan, A. (2020, June). CFDNet: A deep learning-based accelerator for fluid simulations. In *Proceedings of the 34th ACM international conference on supercomputing* (pp. 1-12). DOI: 10.1145/3392717.3392772

Pope, S. B. (1999). A perspective on turbulence modelling. In *Modeling complex turbulent flows* (pp. 53–67). Springer Netherlands. DOI: 10.1007/978-94-011-4724-8_5

Poroseva, S. V., Colmenares, F. J. D., & Murman, S. M. (2016). On the accuracy of RANS simulations with DNS data. *Physics of Fluids, 28*(11), 115102. DOI: 10.1063/1.4966639 PMID: 30271109

Rabault, J., Kuchta, M., Jensen, A., Réglade, U., & Cerardi, N. (2019). Artificial neural networks trained through deep reinforcement learning for active flow control.

Sadrehaghighi, Ideen. (2022). Adaptive Meshing.

Shan, T., Tang, W., Dang, X., Li, M., Yang, F., Xu, S., & Wu, J. (2020). Study on a fast solver for Poisson's equation based on deep learning technique. *IEEE Transactions on Antennas and Propagation, 68*(9), 6725–6733. DOI: 10.1109/TAP.2020.2985172

Shi, H. Y., Wang, S., & Wang, P. Y. (2024). From simulation to reality: CFD-ML-driven structural optimization and experimental analysis of thermal plasma reactors. *Journal of Environmental Chemical Engineering, 12*(3), 112998. DOI: 10.1016/j.jece.2024.112998

Spalart, P., & Allmaras, S. (1992, January). A one-equation turbulence model for aerodynamic flows. In 30th aerospace sciences meeting and exhibit (p. 439). DOI: 10.2514/6.1992-439

Spalart, P. R. (2000). Strategies for turbulence modelling and simulations. *International Journal of Heat and Fluid Flow*, *21*(3), 252–263. DOI: 10.1016/S0142-727X(00)00007-2

Tao, R., Tian, F., Huang, H., & Yang, H. (2023). AI-accelerated computational framework for fluid–structure interaction analysis in aerospace applications.

Vinuesa, R., Hosseini, S. M., Hanifi, A., Henningson, D. S., & Schlatter, P. (2017). Pressure-gradient turbulent boundary layers developing around a wing section. *Flow, Turbulence and Combustion*, *99*(3-4), 613–641. DOI: 10.1007/s10494-017-9840-z PMID: 30069159

Wang, B., & Wang, J. (2021). Application of artificial intelligence in computational fluid dynamics. *Industrial & Engineering Chemistry Research*, *60*(7), 2772–2790. DOI: 10.1021/acs.iecr.0c05045

Weller, H. G., Tabor, G., Jasak, H., & Fureby, C. (1998). A tensorial approach to computational continuum mechanics using object-oriented techniques. *Computers in Physics*, *12*(6), 620–631. DOI: 10.1063/1.168744

Wu, J., Xiao, H., Sun, R., & Wang, Q. (2019). Reynolds-averaged Navier–Stokes equations with explicit data-driven Reynolds stress closure can be ill-conditioned. *Journal of Fluid Mechanics*, *869*, 553–586. DOI: 10.1017/jfm.2019.205

Yarlanki, S., Rajendran, B., & Hamann, H. (2012, May). Estimation of turbulence closure coefficients for data centers using machine learning algorithms. In 13th InterSociety Conference on Thermal and Thermomechanical Phenomena in Electronic Systems (pp. 38-42). IEEE. DOI: 10.1109/ITHERM.2012.6231411

Zhang, Y., Liu, J., Shi, S., & Wang, H. (2021). Machine learning-assisted prediction of ice accretion on aircraft surfaces.

Zhu, L., Zhang, W., Kou, J., & Liu, Y. (2019). Machine learning methods for turbulence modelling in subsonic flows around airfoils. *Physics of Fluids*, *31*(1), 015105. DOI: 10.1063/1.5061693

Chapter 7
Investigation of Computer Vision and Machine Learning to Enhance Quality Control Processes in Aerospace Manufacturing:
Innovative Machine Learning Applications in the Aerospace Industry

Dhirendra Patel
https://orcid.org/0000-0003-3308-168X
Amity University, Greater Noida, India

M. L. Azad
Amity University, Greater Noida, India

ABSTRACT

The integration of computer vision and machine learning (ML) in aerospace manufacturing has revolutionized quality control processes by enabling

DOI: 10.4018/979-8-3693-7525-9.ch007

automated defect detection, predictive maintenance, and real-time monitoring. Traditional quality inspection methods often involve manual assessments, Computer vision systems, powered by deep learning algorithms, enhance defect detection accuracy by analysing images and sensor data, Machine learning further optimizes quality control by predicting potential failures through data-driven insights, reducing production downtime, and improving overall efficiency. This paper explores recent advancements in computer vision and ML for aerospace manufacturing, highlighting their impact on defect classification, anomaly detection, and process optimization. The study also discusses challenges such as data availability, model interpretability, and computational resource requirements. By leveraging intelligent automation, aerospace manufacturers can achieve higher reliability, cost efficiency, and compliance with stringent industry standards.

INTRODUCTION

The aerospace industry demands high precision and reliability in manufacturing processes due to the critical nature of its components. Even minor defects in aircraft structures and engine parts can lead to severe safety risks, increased maintenance costs, and operational inefficiencies. Traditional quality control (QC) methods rely on manual inspection, which is labor-intensive, time-consuming, and prone to human error. As aerospace manufacturing processes become more complex, there is a growing need for advanced inspection techniques that enhance accuracy, efficiency, and consistency, (Agarwal *et al.,* 2023).

Computer vision and machine learning (ML) have emerged as transformative technologies in quality control, enabling automated, data-driven decision-making. Computer vision leverages high-resolution imaging, 3D scanning, and advanced signal processing to identify surface defects, geometric deviations, and material inconsistencies with high precision. Meanwhile, ML algorithms, particularly deep learning models, can analyze vast amounts of data to detect patterns, classify defects, and predict potential failures before they occur. This predictive capability helps manufacturers reduce downtime, optimize maintenance schedules, and improve production efficiency. Recent advancements in artificial intelligence (AI) have further enhanced defect detection capabilities, allowing real-time monitoring of production lines using cameras and sensors. Techniques such as convolutional neural networks (CNNs) for image classification, generative adversarial networks (GANs) for anomaly detection, and reinforcement learning for adaptive quality control have demonstrated significant improvements in defect identification and process optimization. Despite these advancements, challenges remain, including the need for large, high-quality datasets, model interpretability, and integration with existing manufacturing systems. This study explores the role of computer vision and ML in aerospace manufacturing quality control, discussing their applications, benefits, challenges, and prospects, (Bao, 2016) (Butler *et al.,* 2021). By leveraging AI-driven automation, aerospace manufacturers can achieve higher reliability, reduce costs, and ensure compliance with stringent industry standards.

Aerospace manufacturing is a highly complex and precision-driven industry where quality control (QC) is paramount to ensure safety, performance, and compliance with stringent regulatory standards. Traditional quality control methods, which rely on human inspection and manual testing, can be time-consuming, error-prone, and inefficient, (Chebbi *et al.,* n.d.) (Deliencourt *et al.,* n.d.) (Donecle, n.d.) (*Barron's,* 2024) (Balani *et al.,* 2021) (Humfeld *et al.,* 2021) (James, 2016). With the advancements in artificial intelligence (AI), particularly in computer vision and machine learning (ML), new opportunities have emerged to revolutionize QC processes. This chapter investigates the role of computer vision and ML in enhancing quality control in aerospace manufacturing, exploring their applications, benefits, challenges, and future trends.as shown by the Figure 1 flow chart AI in aerospace QC. In the highly regulated and safety-critical aerospace industry, quality control is not merely a requirement, it's a cornerstone of operational integrity and passenger safety. Aerospace components must adhere to the highest standards of precision, reliability, and durability due to the extreme conditions they face in operation. Even the smallest defect or deviation from specifications can lead to catastrophic failures, costly recalls, or production delays. As such, the need for robust, accurate, and efficient quality assurance processes is more pressing than ever.

Traditionally, quality control in aerospace manufacturing relies heavily on manual inspection, conventional nondestructive testing (NDT), and rule-based measurement systems. While these methods have served the industry well, they often suffer from inherent limitations including operator fatigue, subjectivity, limited scalability, and the inability to detect certain types of complex or subtle defects. With increasing production volumes and the adoption of advanced materials and manufacturing techniques such as composites and additive manufacturing, these limitations are becoming more pronounced.

Conventional quality control techniques, including manual inspection, coordinate measuring machines (CMMs), and traditional non-destructive testing (NDT) methods like ultrasonic and radiographic testing, have been the backbone of aerospace manufacturing. However, these methods are increasingly challenged by the complexity of modern materials, the miniaturization of components, the rise of intricate manufacturing methods such as additive manufacturing (AM), and the pressure to reduce cycle time and production costs. Human inspectors, while skilled, are subject to fatigue, inconsistencies, and limitations in detecting minute or hidden defects, (James, 2016) (Liu *et al.,* 2025).

With the global push toward Industry 4.0 and smart manufacturing, the aerospace sector is undergoing a technological transformation. At the core of this transformation are computer vision (CV) and machine learning (ML)two fields that have matured significantly due to advancements in computing power, sensor technology, and data science. Together, these technologies are being leveraged to develop intelligent inspection systems capable of performing rapid, high-precision analysis of parts and assemblies.

Computer vision systems use cameras, lighting, and algorithms to interpret visual information from the manufacturing environment. They can detect surface defects such as cracks, scratches, dents, and deformations; perform geometric measurements; verify assembly steps; and guide robotic systems in automated tasks. These systems offer the advantage of consistent inspection criteria, non-contact measurements, and integration with real-time monitoring systems.

Machine learning, particularly deep learning through convolutional neural networks (CNNs), enables these vision systems to improve over time by learning from vast amounts of labeled data. ML models can distinguish between acceptable variations and critical defects, adapt to new part geometries, and even predict failure modes based on historical patterns. These capabilities are especially valuable in aerospace applications where variability is high and product life cycles are long.

The integration of CV and ML technologies into aerospace manufacturing environments provides a pathway to:

- Increase defect detection accuracy and consistency
- Enable real-time feedback for in-process control
- Reduce dependency on manual inspection
- Improve traceability and documentation through automated data collection
- Enhance predictive maintenance and lifecycle management of components

To address these challenges, emerging technologies such as computer vision and machine learning (ML) offer promising solutions. Computer vision enables machines to process and analyze visual information from images or videos, emulating human visual perception with greater consistency and speed. This capability can be harnessed to automate defect detection, dimensional inspection, surface analysis, and part identification. Meanwhile, machine learning, particularly deep learning, allows systems to learn from large datasets, recognize complex patterns, and make intelligent predictions or classifications without being explicitly programmed. In the context of aerospace manufacturing, integrating computer vision and ML has the potential to drastically enhance the speed, accuracy, and adaptability of quality control systems. These technologies can detect surface defects in composite materials, identify micro-cracks in turbine blades, inspect weld quality, verify component assembly, and even predict potential failures through data-driven insights. When deployed effectively, they reduce human error, minimize inspection time, improve traceability, and support real-time decision-making. This investigation aims to explore the current state-of-the-art applications of computer vision and machine learning in aerospace quality control, assess the effectiveness of these techniques compared to traditional methods, and identify the challenges and opportunities involved in their adoption. Specific objectives include as shown by Figure 1 and 2.

Figure 1. Advance quality control in aerospace

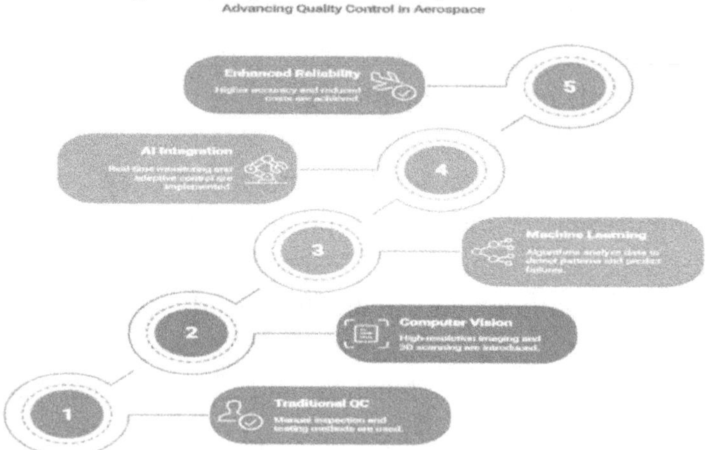

Figure 2. AI in aerospace manufacturing quality control

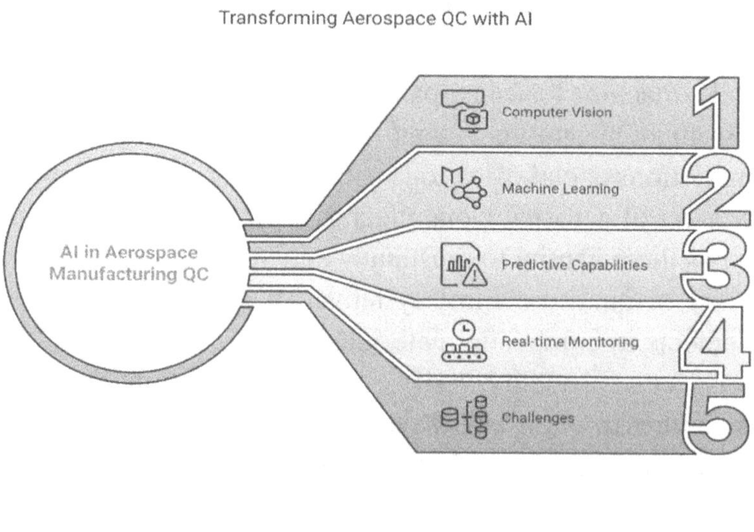

THE ROLE OF QUALITY CONTROL IN AEROSPACE MANUFACTURING

Quality control (QC) is a critical aspect of aerospace manufacturing, ensuring that every component and system meets strict industry standards for safety, reliability, and performance. Given the high cost and safety implications of aerospace components, QC processes must be highly precise, automated, and efficient to minimize risks and maintain compliance with regulatory requirements,

Importance of Quality Control in Aerospace

Aerospace components are subjected to extreme conditions, including high temperatures, intense vibrations, and varying atmospheric pressures. Even a minor defect can compromise structural integrity and lead to catastrophic failures. Therefore, quality control plays a crucial role in

- Ensuring Passenger and Crew Safety: Aircraft components must comply with strict international safety regulations (AS9100, FAA, EASA) to prevent in-flight failures.
- Maintaining Structural Integrity: Aerospace structures, including fuselages, wings, and landing gear, require high durability and defect-free fabrication.
- Enhancing Performance and Reliability: Well-maintained and precisely manufactured components lead to fuel efficiency, extended lifespan, and better aerodynamics.
- Reducing Costs and Wastage: Effective QC prevents production defects, rework, and recalls, saving millions in manufacturing costs, (Gill, 2024) (Kato, 2025) (Elisa, 2024) (Singh & Patel, 2020) (Zamaraev, 1996).

As shown by the diagram in Figure 3.

Figure 3. Improve aerospace quality control

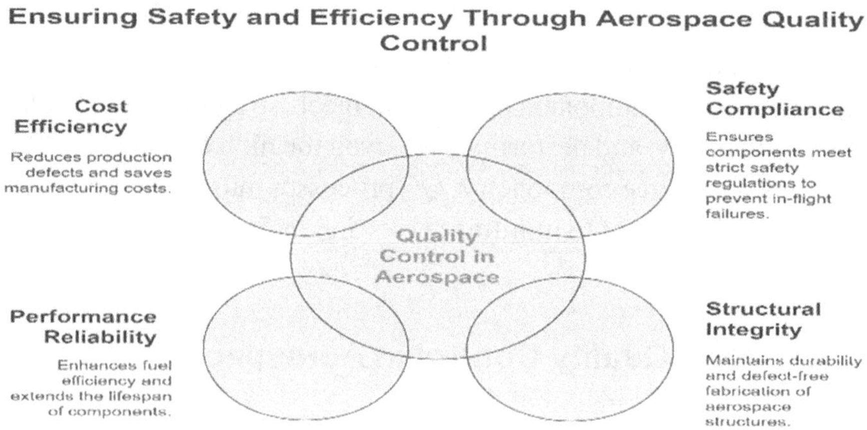

Key Challenges in Aerospace Quality Control

Aerospace manufacturing is highly complex and involves multiple challenges, such as

- Stringent Regulatory Compliance: Meeting standards like AS9100 (Aerospace Quality Management), FAA (Federal Aviation Administration), and EASA (European Union Aviation Safety Agency) require continuous testing and documentation.
- High Precision Requirements: Aerospace parts have micro-level tolerances, meaning even small deviations can lead to component failure.
- Material Complexity: The use of advanced composites, titanium alloys, and lightweight metals requires specialized inspection methods for detecting internal defects.
- Production Volume & Speed: Increasing demand for aircraft necessitates faster and more automated QC processes to keep up with production rates, (Agarwal *et al.*, 2023) (Root, 2024) (Agarwal *et al.*, 2023).

- Cost Considerations: Implementing cutting-edge QC solutions (such as AI-driven inspections) involves high initial costs and integration challenges with legacy manufacturing systems. As shown by the diagram Figure 4.

Figure 4. Aerospace manufacturing in quality

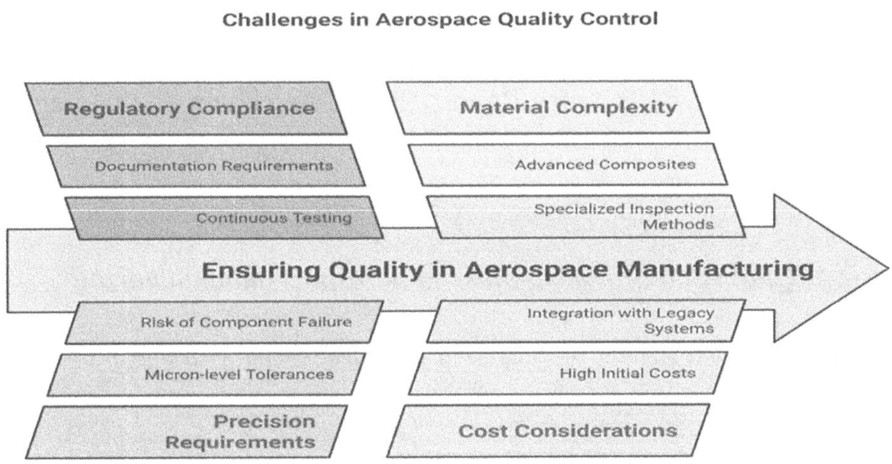

Methods of Quality Control in Aerospace Manufacturing

- To address these challenges, aerospace manufacturers employ various QC techniques, including:
- Manual & Visual Inspection: Traditional method relying on highly trained inspectors using microscopes, magnifiers, and borescopes.
- Automated Optical Inspection (AOI): Uses high-resolution cameras and AI-driven pattern recognition to detect surface defects, cracks, and misalignments.
- Non-Destructive Testing (NDT): Methods such as ultrasonic testing (UT), X-ray, and computed tomography (CT) to identify internal defects without damaging components.

- Coordinate Measuring Machines (CMM): Provides high-precision 3D measurements to verify the dimensions of critical parts like turbine blades and fuselage panels.
- 3D Scanning & Metrology: Laser-based 3D scanners compare manufactured parts with CAD models for dimensional accuracy.
- AI and Machine Learning in QC: AI-powered computer vision systems analyze large volumes of data in real-time, improving efficiency, accuracy, and defect detection rates, (Butler et al., 2021) (Patel, Mishra, & Nabeel, 2022) (Patel, Sharma, & Mishra, 2021) (Chebbi *et al.,* 2021) (Miranda, Larnier, & Claybrought, 2020) (Poudel *et al.,* 2024) (PwC Industry Report, 2023). As shown by the diagram Figure 5.

Figure 5. Quality control techniques in aerospace manufacturing

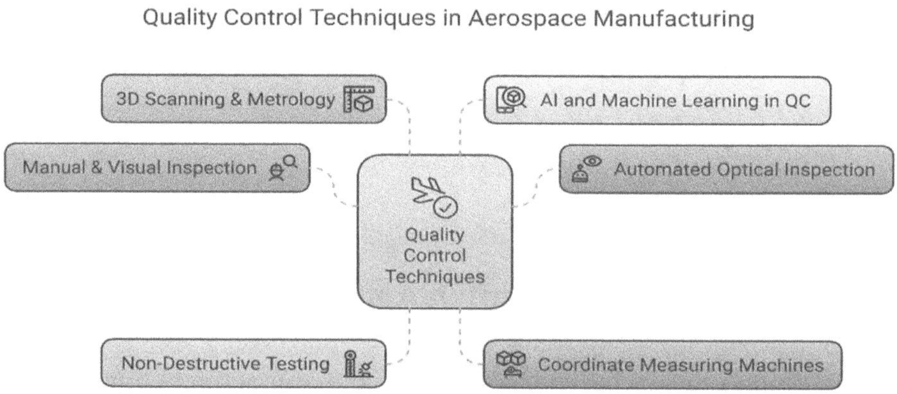

Role of AI, Computer Vision, and Machine Learning in QC

With advancements in computer vision and machine learning, traditional QC methods are being replaced with automated, AI-driven solutions that offer:

- Higher Accuracy: AI detects microscopic defects that humans might miss.
- Faster Inspection: AI-powered vision systems scan thousands of components per second.
- Predictive Maintenance: ML algorithms forecast potential failures before they happen. Cost Reduction: Minimizes material waste, rework, and labor costs, (Broadline Components, 2025) (Patel *et al.*, 2022) Bakirci & Bayraktar, 2025) (Patel, 2023) (Kato, 2025) (McKinsey & Company Report, 2023) (Barlow & Peterson, 2022) (Patel, Varma, & Khan, 2022) (Chatterjee & Gupta, 2023) (Roberts, 2024). As shown by the diagram Figure 6.

Figure 6. AI advancements in quality control

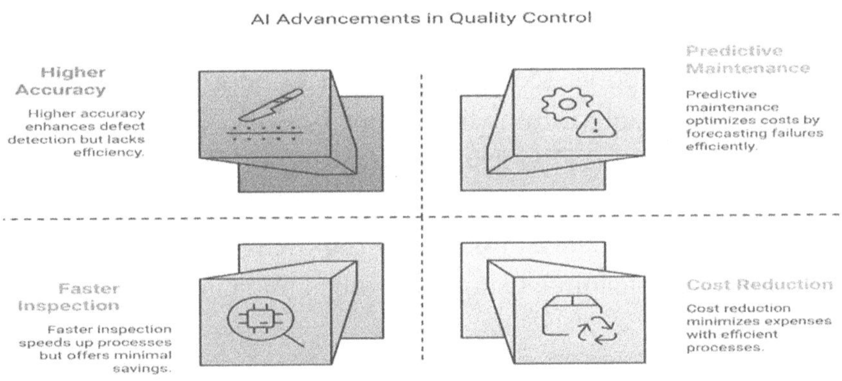

INTRODUCTION TO COMPUTER VISION AND MACHINE LEARNING

Computer Vision

Computer vision is a specialized field of artificial intelligence (AI) that enables machines to interpret, analyze, and understand visual data from the real world. In aerospace manufacturing, computer vision plays a crucial role in quality control by automating the detection of defects, ensuring precision, and improving manufacturing efficiency. It involves several key techniques, (Shah, 2021) (Kumar, Patel, & Singh, 2023) (Wong, 2023) (Core Business Solutions, 2025) (FAA, 2024).

Techniques in Computer Vision

1. Image Processing:
    - Edge Detection: Identifies boundaries between objects and defects using algorithms like Canny or Sobel filters.
    - Segmentation: Divides an image into different regions to isolate objects or defects.
    - Filtering: Enhances images by removing noise or highlighting important features using techniques like Gaussian and median filtering.
2. Object Detection and Recognition:
    - Used to locate and classify parts or defects in aerospace components.
    - Techniques like YOLO (You Only Look Once) and SSD (Single Shot MultiBox Detector) can rapidly identify multiple defects in real time.
3. Defect Detection and Classification:
    - Identifies surface defects, cracks, corrosion, or misalignments in aircraft components.
    - Advanced methods use AI models trained on defect datasets to classify types and severity levels.
4. 3D Reconstruction and Measurement:

- Generates 3D models of components using stereo vision, LiDAR, or structured light scanning.
- Ensures dimensional accuracy by comparing manufactured parts to CAD models.

As shown by the diagram Figure 7.

Figure 7. Computer vision techniques in aerospace manufacturing

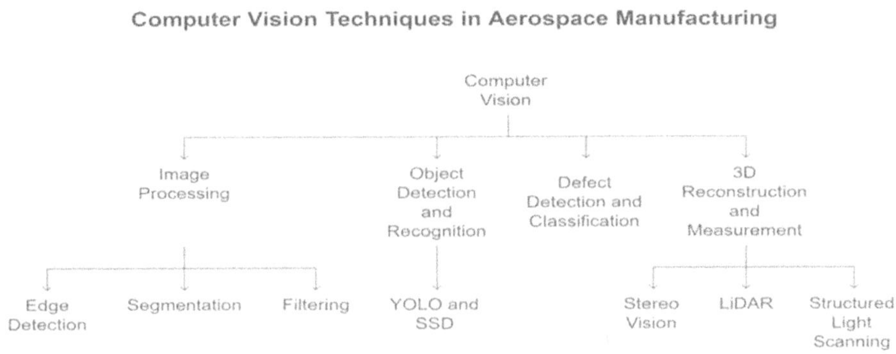

Machine Learning

Machine learning (ML) is a subset of AI that enables systems to learn patterns from data and make predictions or decisions without explicit programming. In aerospace manufacturing, ML enhances quality control by improving defect detection accuracy, reducing human error, and enabling predictive maintenance, (FAA, 2024) (Easa, 2023) (ISO, 2022) (SAE ARP, 2024) (Patel *et al.*, 2017) (Journal of Non-Destructive Testing, 2023) (IEEE Transations on Industrial Informatics, 2024) (Omdena, 2023) (Aerospace Engineering review, Patel, 2024) (Reliability Engineering & System Safety, 2023).

Types of Machine Learning Techniques in Quality Control

1. Supervised Learning:

- o Uses labeled datasets where defects are manually classified.
- o Models like Support Vector Machines (SVM) and Random Forests are trained to distinguish between defective and non-defective parts.
- o Example: Training an ML model with thousands of labeled crack images to classify cracks in new components.

2. Unsupervised Learning:
   - o Does not require labeled data; instead, it detects patterns and anomalies based on statistical distributions.
   - o Methods like k-means clustering or autoencoders identify unusual defects without predefined labels.
   - o Example: Detecting unknown defects in turbine blades by comparing new images with standard manufacturing patterns.
3. Deep Learning (DL):
   - o A specialized branch of ML that uses artificial neural networks, particularly Convolutional Neural Networks (CNNs), for complex image analysis.
   - o CNNs can automatically extract features from images without manual feature engineering.
   - o Example: Using ResNet or VGG-16 models to analyze high-resolution images of aircraft fuselage for cracks and corrosion.

As shown by the diagram Figure 8 Machine learning techniques in aerospace quality control.

Figure 8. Machine learning techniques in aerospace quality control

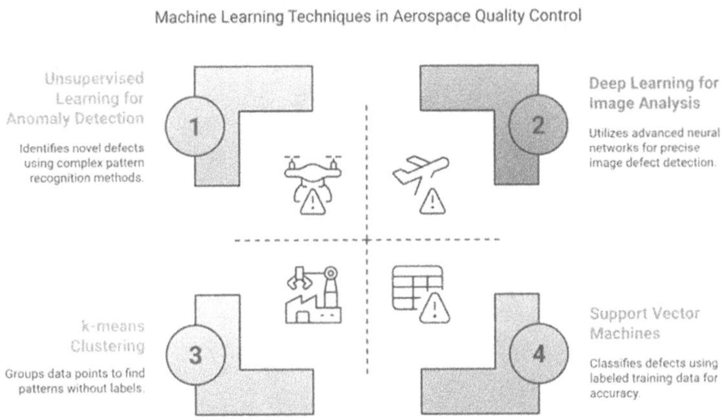

APPLICATIONS OF COMPUTER VISION AND MACHINE LEARNING IN AEROSPACE QC

Automated Defect Detection Automated Defect Detection in Aerospace Using Computer Vision and Machine Learning

Automated defect detection in aerospace manufacturing and maintenance is critical for ensuring safety, reliability, and compliance with stringent quality standards. AI-powered computer vision and machine learning (ML) models enhance defect identification in aircraft components, reducing human error and improving efficiency (Biship, 2006) (Goodfellow, Bengio, & Courville, 2016) (LeCun, Bengio, & Hinton, 2015) (Zhang, he, & Wei, 2020).

- **Surface defect detection:** AI-powered vision systems can identify cracks, scratches, and deformations in metal and composite materials.

- **Corrosion monitoring:** Real-time assessment of corrosion in aircraft parts using hyperspectral imaging.
- **Weld and joint inspection:** Ensuring structural integrity in welded and riveted components.

As shown by Figure 9 Enhancing Aerospace safety through automated defect detection.

Figure 9. Enhancing aerospace safety through automated detection

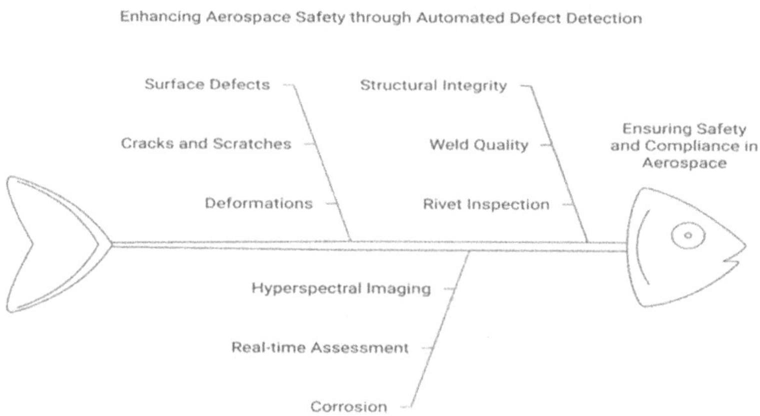

Dimensional Inspection and Precision Measurement

Dimensional inspection and precision measurement are critical in aerospace manufacturing and maintenance to ensure components meet tight tolerances and conform to specifications. Computer vision and machine learning (ML) improve accuracy, reduce human error, and enhance efficiency in these processes, (Beltrán-González, Smith, & Lee, 2020) (Dasari, Nguyen, & Kumar, 2020) (Finney, Johnson, & NASA, 2019) (Patel, 2025) (Graham-Knight *et al.,* 2021) (Maghami, Salehi, & Khoshdarregi, 2021) (Patel, 2018) (Zhang, Li, & Zhao, 2021) (Eagle Technologies, 2022) (Ultralytics, 2022) (ComplainceQuest, 2022).

- **3D scanning & metrology:** Ensuring components meet design tolerances using laser-based 3D scanners.
- **Automated coordinate measurement machines (CMMs):** Enhancing precision in geometric verification.

As shown by the Figure 10 Enhancing Aerospace Precision with Advanced Inspection Technologies.

Figure 10. Enhancing aerospace precision with advanced inspection technologies

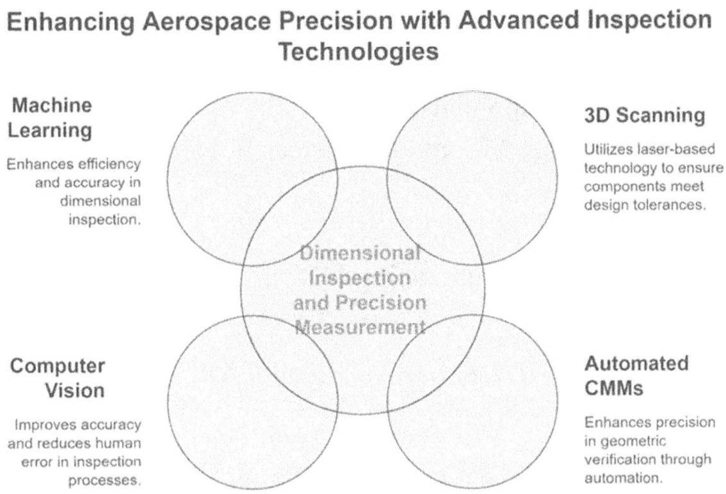

Non-Destructive Testing (NDT) Enhancement Using Computer Vision and Machine Learning in Aerospace

Non-Destructive Testing (NDT) is critical in aerospace for ensuring the integrity and safety of aircraft components without causing damage. Computer vision and machine learning (ML) significantly enhance NDT by improving detection, reducing human error, and increasing efficiency in inspections, (InovarTech, 2022) (ResearchGate, 2022) (ScitePress, 2024) (Lee & Wong, 2022) (Chen & Zhao, 2021) (Nguyen & Tran, 2020) (Martinez & Gonzales, 2023) (Singh & Patel, 2021).

Technologies in AI-enhanced NDT

- **Deep Learning (CNNs, Transformers)** – Automates flaw detection in X-ray, ultrasound, and thermographic images.
- **Computer Vision Algorithms** – Identifies cracks, voids, and delamination in real-time.
- **Hyperspectral & Infrared Imaging** – Detects subsurface defects in composites and metal structures.
- **Acoustic Emission & AI Analysis** – Recognizes failure patterns in material stress testing.
- **3D Scanning & LiDAR** – Ensures precise defect localization in structural components.
- **X-ray and CT scan analysis:** AI algorithms analyze internal defects in aircraft components.
- **Ultrasonic testing:** ML models detect hidden flaws in composite structures.

As shown by Figure 11 Enhancing NDT in Aerospace with AI

Figure 11. Enhancing NDT in aerospace with AI

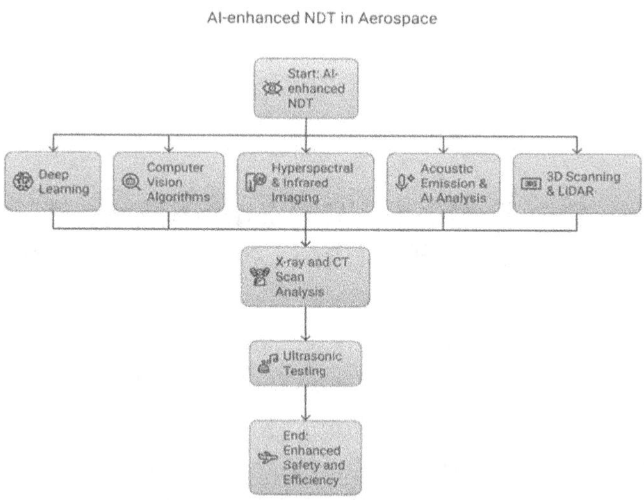

Predictive Maintenance and Process Optimization in Aerospace Using AI

Predictive maintenance (PdM) and process optimization in aerospace leverage computer vision, machine learning (ML), and AI to anticipate failures, reduce downtime, and improve operational efficiency. These technologies help detect anomalies early, optimize maintenance schedules, and enhance manufacturing processes, (Kumar & Sharma, 2020) (Anderson & Brown,2023) (Aust & Pons, 2022) (Beltrán-González, Smith, & Lee, 2020) (Dasari, Nguyen, & Kumar, 2020).

- **Real-time monitoring of production lines** to prevent quality issues before they occur.
- **AI-driven root cause analysis** to reduce rework and waste.

ADVANTAGES OF AI-DRIVEN QUALITY CONTROL

This section highlights how AI enhances precision, efficiency, and reliability. Key points include, (Graham-Knight *et al.*, 2021) (Maghami, Salehi, & Khoshdarregi, 2021) (Meister, Chen, & Wang, 2021) (Sivabalakrishnan, Patel, & Lee, 2020) (Smith & Jones, 2021) (Zhang, Li, & Zhao, 2021).

Increased Accuracy and Consistency

- **Eliminates Human Error** – AI-based inspections detect microscopic defects that human inspectors might miss.
- **Standardized Decision-Making** – Ensures consistent quality assessments across different production batches.
- **Advanced Image Processing** – Computer vision can analyze defects in X-rays, ultrasonic scans, and thermographic images with higher precision.

Faster Inspection and Real-Time Analysis

- **Automated Visual Inspection –** AI analyzes thousands of images in seconds, reducing QC cycle times.
- **Real-Time Defect Detection –** ML models process sensor data instantly, flagging anomalies without delays.
- **Edge AI & IoT Integration –** AI-driven QC systems operate directly on manufacturing lines, reducing bottlenecks.

Predictive Maintenance and Cost Reduction

- **Early Failure Prediction –** AI detects early signs of wear and fatigue before components fail.
- **Optimized Maintenance Schedules** – Reduces unnecessary part replacements, lowering maintenance costs.
- **Minimizes Scrap & Rework –** AI ensures defect-free components, reducing material waste.

Enhanced Safety and Regulatory Compliance

- **Stringent Aerospace Standards –** AI ensures compliance with FAA, EASA, and AS9100 quality regulations.
- **Automated Report Generation –** ML systems document inspection results for traceability and auditing.
- **Comprehensive Data Logging –** AI-based QC provides historical defect tracking for forensic analysis.

As shown by the Figure 12 Benefits of AI-Driven Quality Control

Figure 12. Benefits of AI-driven quality control

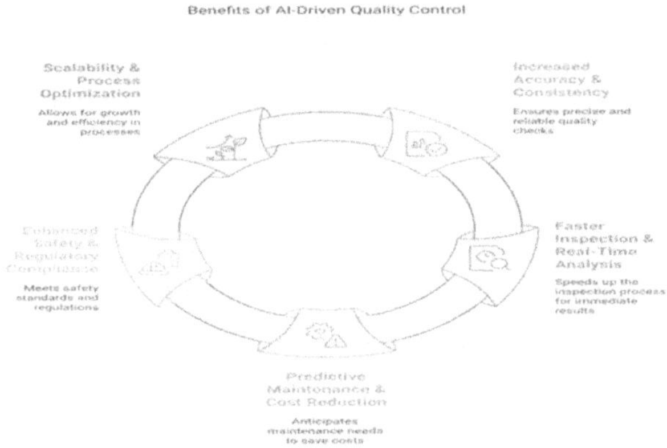

CHALLENGES AND LIMITATIONS

While AI-driven quality control in aerospace manufacturing presents significant advantages, several challenges and limitations must be addressed to ensure seamless integration and efficiency. These challenges span technical, regulatory, financial, and workforce-related issues.

High Implementation Costs

- **Infrastructure Investment**: Deploying AI-powered quality control systems requires significant investments in hardware (high-resolution cameras, sensors, GPUs) and software (machine learning algorithms, cloud computing platforms).
- **Maintenance and Upgrades**: AI models require continuous updates and recalibration, adding to long-term operational costs.
- **Return on Investment (ROI) Concerns**: The high initial costs and uncertain ROI may make companies hesitant to fully embrace AI-driven QC solutions.

Complexity in AI Model Training and Data Requirements

- **Data Scarcity and Quality**: AI models rely on vast amounts of high-quality, labeled training data. Obtaining sufficient defect images, especially for rare failure cases, is challenging.
- **Material and Design Variability**: Aerospace components come in diverse materials (metals, composites, ceramics), making it difficult to generalize AI models across all manufacturing scenarios.
- **Edge Cases and False Positives/Negatives**: AI systems may struggle with unique or unexpected defect patterns, leading to false detections that impact production efficiency.

Regulatory and Compliance Barriers

- **Stringent Aerospace Standards**: AI-driven quality control must comply with strict industry regulations such as FAA (Federal Aviation Administration), EASA (European Union Aviation Safety Agency), and AS9100 standards.
- **Explainability and Transparency**: Machine learning models, especially deep learning networks, often function as "black boxes," making it difficult to explain their decisions to regulators.
- **Validation and Certification**: AI-based systems require extensive testing and validation before they can be approved for safety-critical inspections in aerospace manufacturing.

Integration with Legacy Systems

- **Compatibility Issues**: Many aerospace manufacturers use legacy QC systems that are not AI-compatible, making integration complex and costly.
- **Data Silos and Connectivity**: AI-driven QC requires real-time data sharing across various production units, but disconnected legacy systems create data bottlenecks.
- **Cybersecurity Concerns**: AI-powered quality control systems involve cloud-based storage and real-time data analysis, which increases exposure to cyber threats and requires robust security measures.

Workforce Adaptation and Skill Gaps

- **Resistance to Automation**: Employees may resist AI adoption due to job security concerns, even though AI is intended to augment human capabilities rather than replace them.
- **Lack of AI Expertise**: Implementing AI-driven QC requires expertise in machine learning, computer vision, and industrial automation, which may not be readily available in traditional aerospace manufacturing environments.
- **Ongoing Training Requirements**: Engineers and quality control teams must undergo continuous training to effectively operate and interpret AI-powered inspection results.

Limitations in Real-Time Processing and Adaptability

- **Processing Speed vs. Accuracy Trade-off**: High-resolution image processing and real-time defect detection require significant computational power, which may slow down production workflows.
- **Environmental Variability**: Factors like lighting conditions, temperature fluctuations, and dust can affect the performance of AI-based vision systems, requiring adaptive calibration.
- **Limited Generalization**: AI models trained on specific aircraft components or materials may not perform well when applied to new designs or manufacturing techniques, necessitating retraining and additional data collection.

FUTURE TRENDS AND INNOVATIONS

AI-Driven Predictive Quality Control

- Advanced machine learning models will predict defects before they occur, reducing scrap and rework.
- Real-time monitoring of production lines using AI will enable adaptive quality control.

Edge AI and Real-Time Processing

- AI models will increasingly be deployed on edge devices, enabling faster defect detection with minimal latency.
- On-device processing will enhance security and reduce reliance on cloud-based systems.

Integration of Digital Twins

- Digital twin technology will create real-time virtual replicas of manufacturing processes, optimizing quality control through AI simulations.
- Predictive analytics will identify weak points in production, improving defect prevention.

AI-Augmented Non-Destructive Testing (NDT)

- AI will enhance NDT techniques such as ultrasonic, thermographic, and X-ray inspections.
- Automated defect recognition will improve the accuracy and efficiency of inspections.

Human-AI Collaboration in Quality Control

- AI-powered decision-support systems will assist human inspectors rather than replace them.
- Augmented reality (AR) and AI-driven guidance will help QC teams analyze defects more effectively.

Sustainable and Green Manufacturing with AI

- AI will optimize resource usage, reducing waste and improving sustainability in aerospace manufacturing.
- Energy-efficient AI algorithms will contribute to lower carbon footprints in production processes.

Enhanced Cybersecurity for AI-Driven QC Systems

- AI-driven cybersecurity solutions will protect quality control systems from cyber threats.
- Blockchain technology may be integrated to enhance traceability and data integrity.

CONCLUSION

The integration of computer vision and machine learning in aerospace quality control offers a transformative leap in manufacturing accuracy, efficiency, and safety. Despite challenges, ongoing research and technological advancements will continue to refine AI-driven quality control, ensuring that aerospace manufacturing remains at the forefront of innovation and reliability. Despite these challenges, advancements in AI and computer vision continue to push the boundaries of quality control in aerospace manufacturing. Addressing these limitations through better data management, regulatory collaboration, AI model interpretability, and workforce upskilling will be key to unlocking the full potential of AI-driven quality assurance. The gradual adoption of AI in quality control, coupled with ongoing research and development, will help aerospace manufacturers overcome existing barriers and move towards more efficient, reliable, and sustainable production processes. AI-driven quality control is revolutionizing aerospace manufacturing by enhancing defect detection, predictive maintenance, and process optimization. While significant challenges remain, continuous advancements in AI, computer vision, and machine learning are rapidly addressing these barriers. The future will see AI seamlessly integrated into quality control frameworks, ensuring more reliable, cost-effective, and high-precision manufacturing processes. As the aerospace industry continues to evolve, embracing AI-driven quality control will be essential in maintaining high safety standards, reducing waste, and improving operational efficiency. Despite these challenges, AI-driven QC is the future of aerospace manufacturing and maintenance. Overcoming these limitations requires strategic investments, regulatory collaboration, and workforce upskilling. AI-driven QC enhances precision, speed, safety, cost savings, and scalability, making it indispensable in modern aerospace manufacturing and maintenance.

REFERENCES

Aerospace Quality Trends in AI-Driven Manufacturing. (2023). *PwC Industry Report.*

Agarwal, A., Ajith, A., Wen, C., ... (2023). Robotic defect inspection with visual and tactile perception for large-scale components. *arXiv preprint arXiv:2309.04590.* DOI: 10.1109/IROS55552.2023.10341590

Agarwal, A., Ajith, A., Wen, C., Stryzheus, V., Miller, B., Chen, M., Johnson, M. K., Rincon, J. L. S., Rosca, J., & Yuan, W. (2023). Robotic defect inspection with visual and tactile perception for large-scale components. *arXiv preprint arXiv:2309.04590* DOI: 10.1109/IROS55552.2023.10341590

Agarwal, A., Ajith, A., Wen, C., Stryzheus, V., Miller, B., Chen, M., Johnson, M. K., Rincon, J. L. S., Rosca, J., & Yuan, W. (2023). Robotic defect inspection with visual and tactile perception for large-scale components. *arXiv preprint arXiv:2309.04590.* DOI: 10.1109/IROS55552.2023.10341590

AI-Augmented NDT in Aerospace: A Review. (2023). *Journal of Non-Destructive Testing,45*(2), 89-105.

AI-Based Predictive Maintenance for Aircraft Components. (2023).. . *Reliability Engineering & System Safety, 95*(4), 460–478.

AI in Aerospace: Opportunities and Challenges. (2023). *McKinsey & Company Report.*

Anderson, M., & Brown, C. (2023). Integration of computer vision systems in aircraft assembly verification. *Aviation Safety Journal, 27*(4), 678–689.

Aust, T., & Pons, D. (2022). Automated defect detection on engine blades: A comparative study. *Journal of Aerospace Engineering, 36*(4), 1–15.

Aust, T., & Pons, D. (2022). Automated defect detection on engine blades: A comparative study. *Journal of Aerospace Engineering, 36*(4), 1–15.

Bakirci, M., & Bayraktar, I. (2025). *AI-Driven Micro Defect Detection for Aerospace.* Neural Concept Research Brief.

Balani, R. S., Patel, D., Barthwal, S., & Arun, J. (2021). Fabrication of Air Conditioning System Using the Engine Exhaust Gas. *Advances in Interdisciplinary Research in Engineering and Business Management*, 369-378.

. Bao, E. (2016). Drones: Donecle opère y compris en indoor, des drones pour inspecter les impacts foudre sur un avion. *L'Usine Nouvelle*.

Barlow, C., & Peterson, R. (2022). *Artificial Intelligence in Aerospace: Applications and Challenges*. Springer.

Beltrán-González, M., Smith, J., & Lee, K. (2020). Machine learning-based visual inspection system for aerospace component defect detection. *International Journal of Advanced Manufacturing Technology, 110*(9-12), 2801–2813.

Beltrán-González, M., Smith, J., & Lee, K. (2020). Machine learning-based visual inspection system for aerospace component defect detection. *International Journal of Advanced Manufacturing Technology, 110*(9-12), 2801–2813.

Bishop, C. M. (2006). *Pattern recognition and machine learning*. Springer.

Butler, G. A., Gu, D., Humfeld, K. D., ... (2021). A machine learning framework for real-time inverse modeling and multi-objective process optimization. *arXiv preprint arXiv:2104.11342*.

Butler, G. A., Gu, D., Humfeld, K. D., Nelson, K., & Zobeiry, N. (2021). A machine learning framework for real-time inverse modeling and multi-objective process optimization of composites for active manufacturing control. *arXiv preprint arXiv:2104.11342*.

Chatterjee, A., & Gupta, S. (2023). *Machine Learning for Quality Control in Aerospace Engineering*. CRC Press.

Chebbi, J., Defaÿ, F., Brière, Y., & Deruaz-Pepin, A. (2020, April). Novel model-based control mixing strategy for a coaxial push-pull multirotor. *IEEE Robotics and Automation Letters*, *5*(2), 485–491. DOI: 10.1109/LRA.2019.2963652

Chebbi, J., Defaÿ, F., Brière, Y., & Deruaz-Pepin, A. (2021). Novel model-based control mixing strategy for a coaxial push-pull multirotor. *IEEE Robotics and Automation Letters.*

Chen, L., & Zhao, Q. (2021). Deep learning applications in aerospace manufacturing quality control. *Journal of Intelligent Systems*, *28*(4), 456–467.

ComplianceQuest. (2022). AI in quality management solutions for the aerospace industry

Components, B. (2025). AI in Quality Control and Compliance in Aerospace. *Broadline Components White Paper.*

Components, B. (n.d.). AI for quality control and compliance in aerospace sourcing. Broadline Components., Retrieved March 28, 2025, from.

Concept, N. (2025). Aerospace parts manufacturing and AI: The efficiency guide. Neural Concept., Retrieved March 28, 2025, from.

Core Business Solutions. (2025). *AS9100 Aerospace Quality Management System. International Aerospace Quality Group*. IAQG.

Dasari, A., Nguyen, P., & Kumar, S. (2020). Quality assurance in additive manufacturing: Defect classification using random forest algorithms. *Additive Manufacturing*, *34*, 101234.

Deep Learning for Visual Inspection in Aerospace. (2024)... *IEEE Transactions on Industrial Informatics*, *60*(1), 25–40.

Deliencourt, B., Giraud, M., Miranda, J., Larnier, S., & Herbulot, A. (n.d.). *Génération photo-réaliste de défauts sur des images de surfaces extérieures d'avions acquises par un drone autonome*. ORASIS, Journées Francophones des Jeunes Chercheurs en Vision Par Ordinateur.

Donecle. (n.d.). In *Wikipedia*. Retrieved March 28, 2025

Eagle Technologies. (2022). 5 applications for advanced vision systems in aerospace.

EASA Guidelines on AI-Based Quality Inspection. (2023). *European Union Aviation Safety Agency*. EASA.

. Elisa (2024). AI-driven quality control: Defect detection in manufacturing. *Elisa IndustrIQ*. Retrieved March 28, 2025

FAA AI Integration for Aerospace Safety. (2024). *Federal Aviation Administration*. FAA.

. Finney, K., Johnson, L., & NASA Ames Research Center. (2019). Inspection of multi-material additive manufacturing components: A collaborative approach. *NASA Technical Reports*, 2019-123456.

. Finney, K., Johnson, L., & NASA Ames Research Center. (2019). Inspection of multi-material additive manufacturing components: A collaborative approach. *NASA Technical Reports*, 2019-123456.

Gill, N. S. (2024). AI-driven quality control: Revolutionizing manufacturing standards. Xenon Stack., Retrieved March 28, 2025, from.

Goodfellow, I., Bengio, Y., & Courville, A. (2016). *Deep learning*. MIT Press.

Graham-Knight, B., Körber, M., Djavadifar, A., & Najjaran, H. (2021). Automated visual detection of geometrical defects in composite manufacturing processes using deep convolutional neural networks. *Journal of Intelligent Manufacturing*, *32*(5), 1235–1248.

Graham-Knight, B., Körber, M., Djavadifar, A., & Najjaran, H. (2021). Automated visual detection of geometrical defects in composite manufacturing processes using deep convolutional neural networks. *Journal of Intelligent Manufacturing*, *32*(5), 1235–1248.

Hastie, T., Tibshirani, R., & Friedman, J. (2009). *The elements of statistical learning: Data mining, inference, and prediction*. Springer. DOI: 10.1007/978-0-387-84858-7

Humfeld, K. D., Gu, D., Butler, G. A., Nelson, K., & Zobeiry, N. (2021). A machine learning framework for real-time inverse modeling and multi-objective process optimization of composites for active manufacturing control. *arXiv preprint arXiv:2104.11342*.

InovarTech. (2022). Enhancing worker safety & compliance with AI vision in manufacturing.

ISO 9001: Quality Management Systems for Aerospace. (2022). *International Organization for Standardization (ISO).*

. James, O. (2016). Les drones de Donecle auscultent les avions. *L'Usine Nouvelle.*

Kato, Y. (2025). Balancing Cost and Quality in. Aerospace Manufacturing with AI. *Quality*, *64*(4), 32–32.

Kumar, R., & Sharma, P. (2020). Predictive maintenance in aerospace engineering through machine learning. *Journal of Aerospace Operations*, *8*(2), 123–135.

Kumar, S., Patel, D., & Singh, R. P. (2023). *Thermodynamic Radiators: Principles*. Applications, and Performance Analysis.

LeCun, Y., Bengio, Y., & Hinton, G. (2015). Deep learning. *Nature*, *521*(7553), 436–444. DOI: 10.1038/nature14539 PMID: 26017442

Lee, K., & Wong, M. (2022). Computer vision techniques for real-time aircraft inspection. *International Journal of Aviation Engineering*, *39*(2), 123–134.

Liu, B., Cui, Z., Hu, S., Li, X., Lin, H., & Zhang, Z. (2025). LLM evaluation based on aerospace manufacturing expertise: Automated generation and multi-model question answering. *arXiv preprint arXiv:2501.17183*

Maghami, A., Salehi, M., & Khoshdarregi, M. (2021). Automated vision-based inspection of drilled CFRP composites using multi-light imaging and deep learning. *Journal of Manufacturing Processes*, *64*, 123–134.

Martinez, R., & Gonzalez, L. (2023). Enhancing defect detection in aerospace composites using convolutional neural networks. *Composite Structures*, *250*, 112345.

Meister, R., Chen, Y., & Wang, Z. (2021). Integrating CNN and SVM models for enhanced visual inspection in aerospace applications. *IEEE Transactions on Industrial Informatics*, *17*(6), 4321–4330.

Meister, R., Chen, Y., & Wang, Z. (2021). Integrating CNN and SVM models for enhanced visual inspection in aerospace applications. *IEEE Transactions on Industrial Informatics*, *17*(6), 4321–4330.

Miranda, J., Larnier, S., & Claybrought, M. (2020). AI-based defect detection for aircraft maintenance. *Journal of Aerospace Engineering, 37*(4), 56–72.

Miranda, J., Larnier, S., & Claybrought, M. (n.d.). Caractérisation d'objets sur des images acquises par drone recherchant des défauts pour la maintenance aéronautique. *Actes de la conférence Reconnaissance de Formes et Intelligence Artificielle.*

Mohammadi, A., Al-Fuqaha, A., Sorour, S., & Guizani, M. (2018). Deep learning for IoT big data and streaming analytics: A survey. *IEEE Communications Surveys and Tutorials, 20*(4), 2923–2960. DOI: 10.1109/COMST.2018.2844341

Nguyen, T., & Tran, P. (2020). Automated visual inspection systems in aircraft maintenance. *Aerospace Maintenance Journal, 15*(1), 89–99.

Omdena (2023). Edge AI and Real-Time Defect Detection. *International Journal of Manufacturing Science, 12*(3), 210-230.

Patel, D. (2023). Exploring the Frontiers of Microfluidics: Challenges and Future Prospects. *Advances in MEMS and Microfluidic Systems*, 11-31.

Patel, D. (2024). Sustainable Renewable Energy Sources and Emerging Technologies. *Optimization Techniques for Hybrid Power Systems: Renewable Energy, Electric Vehicles, and Smart Grid*, 343-361.

Patel, D. (2025). Emerging sustainable nanomaterials and their applications and future scope. In *Advances in Sustainable Materials* (pp. 107–135). Elsevier. DOI: 10.1016/B978-0-443-13849-2.00005-3

Patel, D. (2025). Emerging Sustainable Materials to Improve Green Energy: Environmental Applications and Future Scope. In *Innovations in Energy Efficient Construction Through Sustainable Materials* (pp. 65-82). IGI Global.

Patel, D., & Mishra, A. (2023). Hybrid sustainable nanomaterials using for nanofluids of advance applications and challenges of future scope.

Patel, D., Mishra, A., & Nabeel, M. (2022, February). Heat transfer characteristics of nanofluids of silicon oxides (SiO2) with conventional fluid. In *2022 2nd International Conference on Innovative Practices in Technology and Management (ICIPTM)* (Vol. 2, pp. 420-423). IEEE.

Patel, D., Sharma, A., & Mishra, A. (2021, November). Study of convective heat transfer characteristics of nano fluids in circular tube. In *2021 International Conference on Technological Advancements and Innovations (ICTAI)* (pp. 264-267). IEEE. DOI: 10.1109/ICTAI53825.2021.9673432

Patel, D., Singh, A. K., Sarda, C., & Raj, R. (2017). Review Six Stroke Engine. [IJCI]. *International Journal on Cybernetics & Informatics*, *6*(2), 53.

Patel, D., Singh, R. P., Rajput, R. S., & Tiwari, P. (2022, December). Thermophyscical Properties And Applications Nanofluids–On Review. In *2022 5th International Conference on Contemporary Computing and Informatics (IC3I)* (pp. 1324-1328). IEEE. DOI: 10.1109/IC3I56241.2022.10072842

Patel, D., Varma, I. P., & Khan, F. A. (2022). A Review Advanced Vehicle with Automatic Pneumatic Bumper System using Two Cylinder. *Int. Res. J. Eng. Tech*, *9*, 264–267.

Patel, D. K. (2018). Evaluation of Perforance of IC Engine using Alternate Fuel.

Patil, A. (2023). The good, the bad, and the awful of AI in aerospace. Aerospace Manufacturing and Design., Retrieved March 28, 2025, from.

Poudel, L., Jha, S., Meeker, R., ... (2024). Predictive quality assessment for ultrasonic additive manufacturing with deep learning. *arXiv preprint arXiv:2410.24055.*

Poudel, L., Jha, S., Meeker, R., Phan, D.-N., & Bhowmik, R. (2024). Advanced predictive quality assessment for ultrasonic additive manufacturing with deep learning model. *arXiv preprint arXiv:2410.24055.*

Quest, C. (n.d.). AI in quality management solutions for the aerospace industry. Compliance Quest., Retrieved March 28, 2025, from.

Rajput, R. S., Patel, D., & Singh, R. P. (2023). Applications and Feasibility of Large-Scale Solar-Powered Peltier Refrigeration Systems.

ResearchGate. (2022). A review of machine vision applications in aerospace manufacturing quality inspection.

Roberts, J. (2024). *Computer Vision in Manufacturing: Advanced Defect Detection Techniques*. Wiley.

Root, Al (2024). GE partners with Microsoft to bring new AI tools to its workforce. *Barron's*. Retrieved March

Root, Al (2025). GE partners with Microsoft to bring new AI tools to its workforce. *Barron's*. Retrieved March 28, 2025

SAE ARP 6297: AI-Based Quality Control in Aerospace Manufacturing. (2024). *Society of Automotive Engineers (SAE).*

Schmidhuber, J. (2015). Deep learning in neural networks: An overview. *Neural Networks*, *61*, 85–117. DOI: 10.1016/j.neunet.2014.09.003 PMID: 25462637

. ScitePress. (2024). Advanced AI-based solutions for visual inspection: A systematic literature review.

Shah, A. (2021). *AI and the Future of Manufacturing*. Elsevier.

Singh, A. K., & Patel, D. (2020, January). Optimization of Air Flow Over a Car by Wind Tunnel. In *2020 International Conference on Computation, Automation and Knowledge Management (ICCAKM)* (pp. 1-4). IEEE. DOI: 10.1109/ICCAKM46823.2020.9051457

Singh, V., & Patel, S. (2021). Implementation of AI-driven quality control in aerospace assembly lines. *Journal of Manufacturing Science and Engineering*, *143*(6), 061008.

Sivabalakrishnan, M., Patel, R., & Lee, J. (2020). IoT-based additive manufacturing system for customized production. *Journal of Manufacturing Systems*, *56*, 345–356.

Smith, A., & Jones, B. (2021). High-precision 3D reconstruction for small-to-medium-sized objects utilizing line-structured light scanning: A review. *Optical Engineering (Redondo Beach, Calif.)*, *60*(11), 110901.

Smith, A., & Jones, B. (2021). High-precision 3D reconstruction for small-to-medium-sized objects utilizing line-structured light scanning: A review. *Optical Engineering (Redondo Beach, Calif.)*, *60*(11), 110901.

The Role of Digital Twins in AI-Driven QC. (2024)... *Aerospace Engineering Review*, *55*(6), 310–328.

. Ultralytics. (2022). Computer vision in aircraft quality control and damage detection.

Wang, Y., & Li, J. (2022). Real-time monitoring of aerospace manufacturing processes using machine learning. *International Journal of Advanced Manufacturing Technology*, *120*(9-12), 3456–3467.

Wong, K. (2023). *Data-Driven Quality Control in Aerospace Engineering*. Springer.

Zamaraev, K. I. (1996). Photocatalysis: State of the art and perspectives. *Studies in Surface Science and Catalysis*, *101*, 35–50.

Zhang, T., Li, H., & Zhao, Y. (2021). Accurate and reliable sealant inspection for aircraft fuel tanks based on 3D point cloud. *Journal of Aerospace Engineering*, *34*(6), 04021085.

Zhang, T., Li, H., & Zhao, Y. (2021). Accurate and reliable sealant inspection for aircraft fuel tanks based on 3D point cloud. *Journal of Aerospace Engineering*, *34*(6), 04021085.

Zhang, Y., He, J., & Wei, B. (2020). Machine learning in nondestructive testing: Advances and perspectives. *NDT & E International*, *116*, 102288.

Chapter 8
Retracted Coaching Communication: A Conceptual Framework for Coaching Clients Toward More Emotionally Intelligent Intrapersonal

Dinorah Nieves
https://orcid.org/0009-0008-1283-9870
Rocky Mountain College of Art and Design, USA

ABSTRACT

Coaching has been described as a process of partnering with clients in a thought-provoking/creative way to inspire them toward maximum personal/professional performance. Often, that work comes down to two major tasks: helping the client communicate more effectively with key stakeholders in their lives and helping clients communicate more healthily with themselves. Yet, despite the salience of this statement, few academic pursuits work to ground the coaching process in the art of emotionally intelligent communications of the interpersonal and intrapersonal sort. This chapter proposes a research-based paradigm for theoretically and practically conceptualizing the coaching process in this way. Utilizing a pracademic approach, analyzing case study, and employing the use of valid and reliable scales, this author puts forth an argument for coaching

DOI: 10.4018/979-8-3693-7525-9.ch008

as an effective process in facilitating the improved emotionally intelligent interpersonal and intrapersonal communication of coaching clients.

INTRODUCTION

Coaching has been defined as "a facilitated, dialogic and reflective learning process that aims to grow the individuals or teams' awareness, responsibility, and choice… Partnering with clients in a thought-provoking and creative process that inspires them to maximize their personal and professional potential" (English, Sabatine & Brownell, 2019). As a scholar-practitioner in the field, this author has witnessed the usefulness of this process in a myriad of approaches - within lifestyle coaching, executive coaching, crisis intervention, and long-term strategy planning. For many coaches, the work comes down to two major tasks: helping the client communicate more effectively with key stakeholders in their lives and helping clients communicate more healthily with themselves. This takes the form of a two-pronged approach for improving how clients talk to themselves and others.

Talking To Others

The most common need for clients when talking to others seems to be in crafting and delivering well-balanced messages that are neither passive nor aggressive. A large part of the coaching work with clients is in moving them into effective communication approaches in which they are able to hold others accountable, while also remaining compassionate; able to ask for what they need, without behaving in ways that are entitled; able to build boundaries while also building community. "Communication involves a balance between knowing when to speak, when to listen, when to let others speak, and when to wait" (Caí & Fink, 2012). Helping clients to do this can often involve helping them understand how communication was modeled for them, and where they learned to communicate as they do. It can also be important to help them identify and manage deeper motives for expression. Equally important, is the ability to expose them to alternative tools and techniques that can facilitate more emotionally intelligent interpersonal communication between them and others.

To address this work, we may turn to William Schutz' (1959) Fundamental Interpersonal Relations Orientation, in which he posits that all human beings possess three interpersonal needs that inform their interpersonal expression. They are inclusion, control and affection. Exploring each of these interpersonal needs and their role in leading clients' interpersonal communication with others, can be helpful to building their self-awareness, and strategizing goals for more effectively meeting their needs. Complementary are the skills, tools and techniques offered to us by research on emotional intelligence. "Emotional intelligence includes self-awareness and impulse control, persistence, zeal and self-motivation, empathy and social deftness" (Goleman, 2007). This can help provide clients with tangible tools for shifting their interpersonal communication practices.

Talking To Themselves

Emotional intelligence skills can also help to shift intrapersonal communication. A key challenge across approaches is in helping clients to build their awareness of self-talk, as well as their confidence and tools to shift limiting cognitive scripts into automatic thinking patterns and personal narratives that better serve them. In this way, coaching of any kind is often centered around a process of modifying intrapersonal communication. Many coaching sessions then, are often an exercise in helping clients to explore, rescript and evolve their intrapersonal communication competence.

From a sociological perspective, our self-talk scripts often mirror the messages of our primary agents of socialization: family, community, media, school, etc. They are the compilation and internalization of social expectations, peer pressure, emotional contracts, and other communications in which we have been taught by others how to see ourselves. In this way, the idea of the looking-glass self provides great insights (Cooley, 1922).

From a psychological perspective, there exists a range of theoretical ideas on inner speech. One in particular, has proven influential for theorizing about its development: Vygotsky's Theory of Cognitive Development. In Vygotsky's (1934/1987) theory of cognitive development, inner speech is the outcome of a developmental process, by which things like conversations and interactions with others, sometimes referred to as "linguistically mediated social exchanges" are transformed into an internalized "conversation" with the self. Basically, we sense something from our interactions with people that we internalize and then replay as part of our self-talk. Research also shows that "private speech does not atrophy after the completion of internalization but can persist into adulthood as a valuable self-regulatory and motivational tool" (Vygotsky, 1934/1987). Thus, research in the area of automatic thinking, dialogical self-theory, intrapersonal intelligence and the looking-glass self can help to theoretically examine the inner conversations of coaching clients and the role they play in the coaching process.

Combining these theoretical frameworks around interpersonal and intrapersonal communications can serve to greatly inform and frame the coaching process. Understanding emotional intelligence as a key criteria of improvement in these categories can offer yet a stronger model to utilize in coaching practice. Yet, despite the synchronicity apparent in this postulation, there appears an absence of academic pursuits grounding the coaching process in the art of emotionally intelligent communications of the interpersonal and intrapersonal sort. This chapter proposes a research-based paradigm for theoretically and practically conceptualizing the coaching process as a communications-centered process focused on improving emotional intelligence in the relational and self-talk spaces. This is a process in which the coach helps clients to explore the genesis, details, and impact of their conversations with themselves and with others, while learning tools for becoming more aware of, challenging, and rewriting those internal and external communications that do not serve them.

APPROACH

Defining a Pracademic Approach

Volpe and Chandler (1999) first coined the term "pracademic," to describe "academics who are scholars and teachers in the field of dispute resolution and actually practice what they preach in their university."

While the definition can be seen as field-specific, the call to action was universal. There seemed a need to value those who not only talked the talk but also walked the walk - who could put into practice, in real and concrete ways, on a daily basis, what they learned and taught. Ten years later, Dr. Paul Posner (2009) expanded the term "pracademic," to refer to what he termed "boundary-spanners… who have occupied significant positions as both academics and practitioners… (and who as) adaptable and cross-pressured actors serve the indispensable roles of translating, coordinating and aligning perspectives across multiple constituencies." These explorations are thought to be built on earlier calls made for the advancement of academics/practitioners, by Dick Zody (1977), and others who pushed to traverse the boundary between practitioner and scholar (Kuhn, 2002).

The approach of a pracademic is very much that of both outsider and insider at all times navigating the tension between the knowledge one brings to the moment, and the way the moment contributes to one's base of knowledge. It is a delicate dance of implementing and analyzing - one that works to honor the conversation between theory and application. Much like in any other conversation, the prompted reflection unearths observations and insights not made easily available to a consciousness focused on just one or the other (analysis OR application). That is the reward for careers often spent on the fringe of each professional landscape - straddling the other - the incredible view from the vesica piscis.

This Pracademic

This chapter's conceptual exploration draws on the positionality of the author as a pracademic working professionally as a counselor, coach, and behavioral scientist while serving as full-time faculty in communications and the social & behavioral sciences. This work furthers the trend toward combining reflective professional practice with academic scholarship (Walker, 2010) in an effort to offer on-the-ground insights that both support and challenge theoretic and research-based frameworks. As such, aspects of the author's experience become pertinent to the work in ways that traditionally defy the professional distance often encouraged in academic discourse. In keeping with that understanding, worth mentioning is the author's professional coaching experience - spanning almost two decades, with formal credentialing earned in 2007, suggesting seventeen years of certified coaching at the time of writing. It is also worth noting that the author's professional coaching roles include but are not limited to youth development, community organizing, executive leadership training, broadcast media consultation, and socio-emotional learning groups. Client ages have ranged from 10-90, and have varied in gender, race, religion, and economic class. All work has thus far been based in the United States of America and has been primarily English-speaking. Both in-person and remote modalities have been utilized. Remote sessions have been conducted by phone and by face-to-face meeting software applications. In-person meetings have occurred in schools, homes, offices, and community spaces.

A THEORETICAL FRAMEWORK FOR COACHING CLIENT COMMUNICATION

Interpersonal Communication in Coaching

Interpersonal communication competence has long been studied by interested parties. A helpful description comes to us from Brian Spitzberg (2015), who explained the vitality of interpersonal communication competence in the following way:

Every act and artifact of communication is open to evaluations of its quality, or how well it is accomplished. Because such evaluations involve individual and social judgments of communicative performance, especially in interpersonal contexts, and because virtually all achievements of interpersonal communication depend on performance and subsequent evaluations, a theory of interpersonal communication competence is tantamount to a theory of interpersonal communication. Interpersonal communication competence and social skills are concerned with both the performance of communication and the evaluation of its quality.

Most clients come into coaching with the author as a result of a glitch in exactly this process. In the course of their daily life - at home, at work, in love - evaluations involving individual and social judgments of their communicative performance have left them with more questions than answers. Plainly, they are in a communication breakdown with someone. They find themselves wanting to accomplish a goal and do not know how to enlist the support of others. They find themselves in conflict with a client, colleague, friend or family member and don't know how to resolve it. They feel lost in their professional journey and cannot see their way to securing powerful partnerships that will advance their agenda. They often come seeking not simply a way forward, but a way between… A way to connect more effectively with those in their lives they perceive as obstacles, or out of reach.

Psychiatrist Harry Stack Sullivan (1953) suggested that understanding interpersonal communication was integral to helping individuals who were struggling to function and/or thrive. In direct contrast to the popular thought of psychiatry as the study of what goes on within a person, Sullivan's interpersonal theory of psychiatry invited us to see psychiatry as, "the study of processes that involve or go on between people." In this way Sullivan put interpersonal relationships and their foundation - interpersonal communication - at the center of the helping professions.

While coaching is decidedly different from psychiatry in both training, execution and client-base, there are alignments in psychodynamic and coaching practices that merit extending Sullivan's ideas into the coaching sphere. Overlaps in goals include, "behavior change, enhanced decision making, and better use of resources" (Jordan & Livingstone, 2013). In fact, while coaching does not treat mental illness as psychotherapy and psychiatry do, modern interventions in both coaching and psychodynamic approaches have been noted to draw from similar paradigms, such as mindfulness, motivational interviewing and strengths-based/solution-oriented foci (Jordan & Livingstone, 2013). In this way, one might see how Sullivan's ideas can easily also apply to the world of coaching as it too, often involves studying processes that go on between people.

A systematic meta-synthesis of 110 peer-reviewed qualitative studies, intended to identify client factors and contextual conditions affecting when and how clients engage in effective coaching, (Erdös,de Haan, & Heusinkveld, 2020) found three emerging client factor dimensions: behavior, attitude and emotion, of which behavior was the most extensively studied and was described as "Behaviour reflects competencies and certain inclinations in interpersonal communication…" This indicates the important role that interpersonal communication skills, analysis, and competence play in effective coaching with clients.

Working with clients to improve interpersonal communication competence then, is a key component of effective coaching work. Continuing the work informed by Spitzberg's research (2009), this requires helping clients to understand the appropriateness (grounded in socially acceptable behavior) and effectiveness (helpful in achieving the desired goal) of their communication. This is best done by helping them to be better "motivated to communicate, knowledgeable about communication, and skilled in communicating" (Spitzberg, 2009).

In this way, Spitzberg outlines the makings of an interpersonal communication coaching plan.

- Help clients feel motivated about communicating.
- Help clients better understand the value and process of communication.
- Help clients learn and practice concrete communication skills.
- Help clients process social norms associated with communication in a particular context.
- Help clients strategize ways to maximize the likelihood that their message will be received effectively.

In addition to the "how," it is often important to help clients understand and address the "why" associated with their interpersonal communication challenges. What lies at the heart of some of the struggles they may be experiencing? For this, one might turn to the insights of American psychologist, William C. Schutz (1958) and his theory of "Fundamental Interpersonal Relations Orientations," developed as a systematic approach to the understanding of interpersonal communication, based on interpersonal needs. Schutz believed that interpersonal needs could be divided into three categories: inclusion, control and affection. These served as the fundamental impetus for interpersonal communication - the desire to express and/or receive one or more of the three. Maintaining a satisfactory relationship, social involvement, and belongingness all constitute "inclusion." The need for influence and power are captured by "control." "Affection," on the other hand, centers on the desire for intimacy, closeness, friendship, and love. Successful interpersonal encounters then occur when the interpersonal needs of the participants are met. For this reason, it is worth adding here, to the interpersonal communication competence agenda in coaching:

- Help clients understand why they are communicating.
- Help clients understand the possible desires behind the communications of others.

In doing this work, the objective is to expose what is not working in a client's interpersonal communication habits, to analyze and understand the challenges as they surface, and to explore alternative interpersonal communication strategies with the client that can facilitate the client's growth toward a desired change in behavior and/or circumstance.

Intrapersonal Communication in Coaching

Without fail, the analysis of every interpersonal communication breakdown seems always to lead to conversations about intrapersonal communication breakdown. Intrapersonal communication has been defined as, "the physiological and psychological decoding, processing, storing, and encoding of messages that happen within individuals at conscious and nonconscious levels whenever they communicate with themselves or others for the purposes of defining, maintaining, or developing their social, psychological, or physical selves" (Roberts, 1983).

In earlier years, many scholars were quick to dismiss queries regarding intrapersonal communication as simply thoughts about thinking; suggesting that it was little more than a synonym for cognition (Vocate, 1994). However, emerging research on imagined interactions (Honeycutt & Ford, 2001) and self-statement modification (Dush, Hirt & Schroeder, 1983) among other related constructs, seem to have advanced our understanding of the now popular concept discussed as self-talk (Hardy, 2006).

While self-talk has made a prominent appearance in trending conversations over recent years, it has been a helpful concept in understanding individual purpose and performance for decades. Perhaps one of the most well-regarded theoretical frameworks on self-talk comes to us from Vygotsky.

Vygotsky (1930/1978) explained self-talk, positioning it as an important aspect of individual identity and behavior - suggesting that children internalize social speech. Vygotsky suggested that, in much the same way that we see children ages three and four speaking to themselves, we continue speaking to ourselves indefinitely, only relocating the performance to an internal stage. While Piaget (1923) before him, referred to this inner speech as egocentric and limited, Vygotsky saw it as socially-influenced and character-defining.

This is very much consistent with Sullivan's (1953) view of an individual's self-system as a collection of internalized perceptions, formed by notions of how we are experienced by others. All of this is also reminiscent of Charles Horton Cooley's (1902) ideas of the Looking-Glass Self. According to Cooley, how one sees themselves is largely based on how one imagines they are perceived by another's mind and thus, 'the kind of self-feeling one has is determined by the attitude toward this attributed to that other mind. A social self of this sort might be called the reflected or looking-glass self: Each to each a looking-glass. Reflect the other that doth pass." The Looking-Glass Self, as specified by Cooley, has three fundamental components: the imagination of our appearance (not simply physical) to the other person; the imagination of their judgment about it, and some sort of evaluative feeling. In short, our self-talk mirrors and impacts our self-perception which is influenced by our interpersonal interactions.

In this way, Vygotsky, Sullivan, and Cooley outline the makings of an intrapersonal communication coaching plan:

- Help clients identify how they believe they are perceived by others.
- Help clients better understand the role their ideas of others' perceptions may play in their self-perception.
- Help clients understand their investment in the perceptions of others.

If we are going to be effective communicators in the world, it is important that we understand and analyze this self-perception. A plethora of theoretical models exist to explain and explore our self-perception. One such model looks specifically at the role that our self-perception plays in our communication - inviting us to consider it as a three-part perspective involving: self-concept, self-esteem, and self-presentation (Stone, 2016).

As discussed, our self-concept is often shaped by the perceptions of others. Fully exploring our self-concept, then often involves understanding the ways that it mirrors what we believe others think and feel about us, and what we remember, others thinking and feeling about us as we develop through childhood and into adulthood. Self-esteem, in contrast, is founded on the value and judgment that we attach to aspects of ourselves. It deals in labels of desirable and undesirable. It can mimic societal narrative, familial, criticisms, social expectations, etc. It's about more than who we think we are, it's about what we think that means in the world. Self-discrepancy theory (Higgins, 1989) is helpful here in examining the ways that coaching clients often describe their discontent. Their current experience seems out of alignment with the way they envisioned it, which often comes from a discrepancy between their actual self and their ideal/ought selves - who they would like to be/think they should be. Self-presentation completes the trifecta of self-perception as it relates to this model of communication. Self-presentation is defined as the process of strategically concealing or revealing personal information in order to influence others' perceptions.

These self-perceptions, exposed in the form of self-talk, become the guiding values, beliefs, thoughts and frameworks through which we develop our intrapersonal communication scripts. As these theoretical perspectives make clear, our self-perception has a significant impact on how and what we communicate. There exists a reciprocal relationship, if you will, between the interpersonal communications in which we interact and the intrapersonal communications in which we engage. For this reason, it is worth adding here, to the intrapersonal communication competence agenda in coaching:

- Help clients articulate their self-concept.
- Help clients understand the values impacting their self-esteem.
- Help clients present in ways that are aligned with their evolving selves.

In addition, the important role that self-compassion plays in mitigating negative self-talk and sponsoring positive self-talk (Stutts, 2022) need also be considered when contemplating emotionally intelligent intrapersonal communication.

Research has shown that self-compassion is associated with psychological well-being and is an important protective factor that fosters emotional resilience… or example, higher levels of self-compassion are typically related to greater psychological health as demonstrated by less depression and anxiety and greater happiness and optimism (Raes et al., 2011).

This certainly earns it a place as a final bullet within the intrapersonal communication strategy.

- Help clients learn and practice self-compassion.

For many coaching clients, this forms the bulk of the intrapersonal work needed to tackle the obstacles they bring into coaching - helping them to explore how they conceptualize, value and present themselves - and working with them to unpack that and positively shift that through the language they use in their self-statements.

Studies have shown a positive correlation between the esteem support (verbal and nonverbal aid provided to another individual to enhance how they feel about themselves and their attributes) of coaches and the positive self-talk of coachees (Zourbanos, Hatzigeorgiadis & Theodorakis, 2007). This also reflects the author's knowledge and experience as a pracademic counselor, coach and behavioral scientist working with coaching clients.

EMOTIONAL INTELLIGENCE IN COACHING

Psychologist Howard Gardner (1983) proposed a theory of Multiple Intelligences (MI), suggesting the existence of varied categories of "smart." These took the form of linguistic, logical, spatial, bodily–kinesthetic, musical, interpersonal, intrapersonal, and naturalist. While a research-validated evaluation of the theory of multiple intelligences may not yet be possible (Ferrero, Vadillo, & León, 2021), Gardner's theory does offer a compelling and insightful conceptual framework with which to examine the process of coaching. The author's professional experience would posit that coaching essentially involves the task of helping individuals to improve their interpersonal and intrapersonal intelligence, competence and thus communication. In fact, in many ways, these coalesce in an improvement of emotional intelligence.

Mayer, Salovey & Caruso (2004) describe emotional intelligence as containing five key components: awareness of personal emotions, emotions management, self-motivation, identifying the emotions of others, and handling relationships. Bar-On (1997) before them, discussed a model of emotional intelligence containing five factors. Those were general mood (happiness and optimism), adaptability (problem-solving, reality testing, flexibility), stress management (stress tolerance, impulse control), interpersonal (empathy, interpersonal relationships and social responsibility), and intrapersonal (emotional self-awareness, assertiveness, self-esteem, self-actualization and independence). But perhaps the most popular theory of emotional intelligence was given to us by Daniel Goleman (1998), who defined emotional intelligence as related to self-awareness, self-control, motivation (self-motivation), empathy (awareness and understanding of the emotions of others) and social skills (interpersonal relationships). Goleman expanded on those five domains, further dividing them into 25 different emotional competences (Goleman, 1998).

From the author's perspective, any of these definitions of emotional intelligence might be further distilled to an even simpler bifurcation: intrapersonal communication competence and interpersonal communication competence - how well does one interact with others and how well does one interact with oneself? Author experience suggests that clients often come to the coaching process in search for advice on resolving professional, familial and/or romantic conflicts. Strategy sessions intended to address these issues quickly expose issues with self-awareness, self-compassion, self-confidence and other internal obstacles to employing reasonable conflict resolution techniques. Immediately necessary then, is a two-pronged approach in which the client and coach are constantly addressing both interpersonal and intrapersonal communications as a way of fluidly exploring challenges at both their social and personal levels. Problem-solving gives way to solution-testing which reveals strengths and struggles, intentions and apprehensions, narratives and goals and ultimately fuller truths surrounding the client's internal and relational setbacks.

A visual representation of this model is depicted (see Figure 1) to begin illustrating the paradigm. Conceptually, this model perceives and presents a coaching paradigm in which clients are engaged in a process of exploring and evolving the distinct, yet intertwined roads of interpersonal and intrapersonal communication competence toward increased emotional intelligence.

Figure 1. Toward an EI communications-focused model of coaching

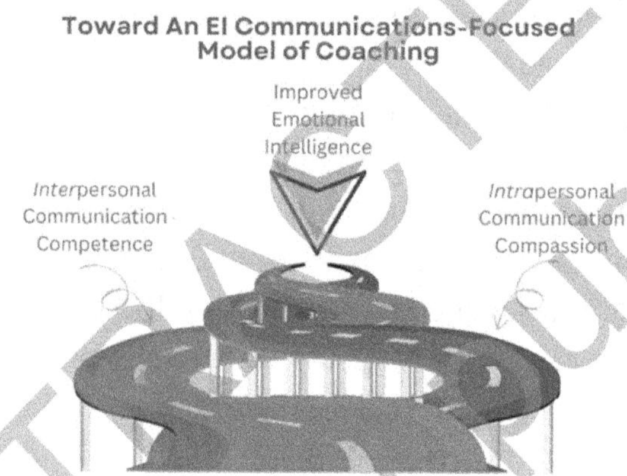

CASE IN POINT

The Case of Michelle

Almost all coaching work then comes down to two separate, yet interrelated questions: what kind of conversations is one having with others, and what kind of conversations is one having with oneself? Take for instance, the case of Michelle - age 50, a straight, African-American woman and CEO of a boutique finance firm with several, multi-million dollar accounts. Michelle sought out executive coaching intending to learn to manage the difficult clients currently in her portfolio. Michelle also requested help processing her service pricing structure and her apprehension around rate increases. Immediately, one can see the duality laid out. Effectively managing difficult clients will likely require that she adjust her interpersonal communications approach. Addressing the apprehension around pricing will likely require that she ask some difficult questions of herself about her perceived self-worth, her confidence to perform, her ideas around abundance and scarcity - all suggesting some profound intrapersonal communication work. Yet each of these also has a "Side B" if you will.

The interpersonal shift required to effectively manage difficult clients will undoubtedly spark intrapersonal communication for Michelle as well - why deal with this client? What are Michelle's negotiables and non-negotiables? Can she assert herself? Can she offer the client some grace? The same holds true for Michelle's second objective - addressing her pricing apprehension. There are various intrapersonal conversations to be sparked as discussed, but there will also be key interpersonal communications that must be effectively navigated to support success in this area. For instance, how will pricing be communicated? Is it done empathetically, clearly? Will she be adaptable? How motivated is she to make and communicate this change? In this way, we can see that what began with two very simple coaching prompts, quickly evolves into two-pronged coaching conversations that explore how Michelle communicates with herself and how she communicates with others. And so, one can easily see how the coaching process expands into a larger process of helping clients unpack and adjust their interpersonal and intrapersonal communication skills, styles and habits.

Michelle attended ninety-four virtual sessions over the course of approximately two years (March of 2022 thru May of 2024). Sessions averaged ninety minutes in duration and totaled approximately one hundred, forty-one hours of coaching by the last structured session. Sessions were discontinued as goals were met, with the understanding that she could revisit the coaching process at any future time should she deem it helpful to do so.

The Data

Information gathered on the efficacy of this coaching approach with Michelle appears in the following three ways:

1. Coaching Session Themes – As explored through coded session notes
2. Client Narrative Feedback – As reported by client list of benefits
3. Client Survey Responses – As completed by client in response to psychometric scales

It should be noted that this work is determined to be exempt by Institutional Review Board (IRB) standards according to the U.S. Department of Health and Human Services (HHS) requirements for exemptions (2018) from their "Basic HHS Policy for Protection of Human Research Subjects." This analysis meets stated HHS criteria for "Exempt Research" qualifying as primarily "research involving benign behavioral interventions in conjunction with the collection of information from an adult subject through verbal or written responses" (Office for Human Research Protections, 2023, subpart d3). Also, "the information obtained is recorded by the investigator in such a manner that the identity of the human subjects cannot readily be ascertained, directly or through identifiers linked to the subjects" (Office for Human Research Protections, 2023, subpart d2i).

Coaching Session Themes

Coaching session notes were retrospectively coded for emerging themes. Fifteen conversations were recurring and central to the work. Taken directly from session notes, shared with the client for accuracy, and reported here, these session notes reveal that a bulk of the time spent between thecoach (the author) and Michelle, was spent working on:

- Practicing boundary-building
- Brainstorming ways to be more assertive
- Challenging ideas of failure and success
- Discussing self-limiting beliefs about responsibility, control, helplessness, codependence, fear, guilt and shame

- Evaluating the continued usefulness of behaviors that "got us here"
- Exploring perfectionism
- Responding effectively and self-compassionately in the face of difficult client conversations
- Processing micro-aggressions by colleagues regarding race and gender
- Processing insecurities and the social scripts that feed them
- Practicing taking accountability without living in blame
- Using questions more than explanations
- Working through imposter phenomenon
- Practicing self-gratitude
- Being honest about not being "alright"
- Challenging a scarcity mindset

Upon further analysis, it is easy to see how these conversations fit into our existing schema. Each either calls for a revised interpersonal communication strategy, or an amended intrapersonal communication script. These conversations can be further distilled into four primary themes:

- Empathy
- Assertiveness
- Self-compassion
- Emotional transparency

All represent an external as well as an internal conflict, which coaching can help the client address by way of interpersonal communication competence and intrapersonal communication compassion.

Client Narrative Feedback

In a post-coaching reflection, Michelle reported (in these exact words) the following concrete professional benefits:

- Developed more clarity and confidence about myself as an entrepreneur and my offerings.

- Incorporated tools and scripts to help me set healthy and fair boundaries with my clients.
- Motivated me from procrasti-planning for unachievable perfection to small, imperfect progress toward my goal.
- Gave me the courage to reach out to prospects and opportunities that I previously thought were unreachable.
- Gave me a sounding board to calibrate my thoughts and feelings about my business and how I wanted to show up in my market.
- Gave me guidance and calibration around the value of my services and my pricing model.
- Helped me spell out and communicate my offers in a clear and concise way that is compelling to my target audience.

Client Survey Responses

In an attempt to capture any shifts in emotional intelligence, interpersonal communication competence or intrapersonal communication patterns, Michelle was asked to use a retrospective pre-post self-assessment approach (Bhanji et al., 2012) to complete three research-validated scales: The Self-Compassion Scale Short Form - SCS-SF (Raes et al., 2011), Spitzberg & Cupach's Interpersonal Communication Competence Self-Assessment (1984) and The Schutte Self Report Emotional Intelligence Test - SSEIT (1998). The author designed a 'Google Sheet' to include three tabs, each adapting one of three scales (SSEIT, ICC, and SCS-SF) intended to measure changes in the coaching model's key components: interpersonal communication competence, intrapersonal communication compassion and emotional intelligence. Change in client's emotional intelligence was assessed by the scientifically valid and reliable instrument "The Schutte Self Report Emotional Intelligence Test (SSEIT)" – a thirty-three item self-inventory designed to measure emotional intelligence across four subscales: emotion perception, utilizing emotions, managing self-relevant emotions and managing others' emotions (Schutte & Malouff, 1998). In terms of assessing interpersonal communication competence, the client completed the eighteen-item "Interpersonal Communication Competence Self-Assessment" (ICC), exploring three basic components of interpersonal communication competence: knowledge, skill, and motivation (Spitzberg

& Cupach, 1984). For the assessment of compassionate self-talk, a 12-item self-report measure was used: the Self-Compassion Scale – Short Form (SCS-SF) – designed to quantify an adult's capacity for self-compassion, meaning their ability to hold their own feelings of suffering with concern, connection and warmth (Res et al., 2011). Michelle's results (see Table 1.) were positive and promising.

Table 1. Client-reported change

Client-Reported Change	Client Score Prior To Coaching	Client Score After Coaching	Difference in Score (Pre to Post)	Increase (Pre to Post)
The Schutte Self Report Emotional Intelligence Test (SSEIT)	116	137	21	18%
Spitzberg & Cupach's Interpersonal Communication Competence Self-Assessment (ICC)	75	82	7	9%
Raes, Pommier, Neff, & Van Gucht's Self-Compassion Scale Short Form (SCS-SF)	31	48	17	55%

Michelle reported an 18% increase in emotional intelligence as measured by The Schutte Self Report Emotional Intelligence Test (SSEIT). This represents a 21-point increase from a score of 116 before beginning the coaching process to 137 reported as her post-coaching score. With regards to interpersonal communication competence, Michelle reported a 9% increase in score (from 75 to 82) as measured by Spitzberg & Cupach's Interpersonal Communication Competence Self-Assessment (ICC). As for compassionate self-talk, Michelle reported an increase of 17 points (55%) from 31 to 48 as measured by Raes, Pommier, Neff, & Van Gucht's Self-Compassion Scale Short Form (SCS-SF).

While the numeric values of these scores cannot be compared due to their difference in measurement style and content, they do all imply notable growth associated with the time spent in coaching, in these key areas: emotional intelligence, interpersonal communication competence and more compassionate self-talk. Michelle's story is one of many with similar objectives, themes and results.

DISCUSSIONS

Overall, the reported experiences of Michelle seem to support the efficacy of the model postulated here. Coaching can be utilized to improve the emotional intelligence of a client's interpersonal and intrapersonal communication to help the client thrive, while addressing the concrete goals that brought them to coaching. Michelle's session notes offer evidence of emotional intelligence work being done within the sessions. Michelle's narrative feedback of benefits taken from the coaching process, confirms professional and personal thriving toward set goals. Moreover, Michelle's responses to survey questions for relevant self-report scales, seem to indicate an improvement in interpersonal communication competence, intrapersonal communication compassion and emotional intelligence.

LIMITATIONS

As a conceptual framework, this chapter is "a description of the way a researcher understands the factors and/or variables that are involved in the study and their relationships to one another... to articulate the concepts under study using relevant literature and to clarify the presumed relationships among those concepts" (Luft et al., 2022). This chapter then, is focused on developing a conceptual framework for coaching clients toward more emotionally intelligent interpersonal and intrapersonal communication. It does this by weaving together the coaching experience of the author with theoretical perspectives and a client case study, to posit a model that can positively impact clients. It does not strive to prove or disprove a hypothesis, to present generalizable research findings, or to argue an irrefutable position.

The data analyzed here (though consistent with the pracademic author and coach's decades of experience) presents the case of a single individual whose coaching experience took place across two years alongside innumerable other influences. This limits the external validity of findings and renders it difficult to conclude, beyond doubt, that the findings are unique to the coaching communications model posited. No control group was utilized to compare and contrast the results of this client. Therefore, it remains unclear to what degree the improvement demonstrated by Michelle can be exclusively attributed to the work described here.

Another limitation involves the scales used to measure improvement. The retrospective decision was made by the author/coach to focus on improved emotional intelligence, increased interpersonal communication competence and increased intrapersonal communication compassion as key components of the coaching work. It may be argued that the client may have different ways of measuring coaching efficacy.

Still, what is clear is that a reported improvement in the interpersonal communication competence, intrapersonal communication compassion and emotional intelligence in the client, seems to coincide (and is associated by the client) with a coaching process that addressed those very themes. Future research on the matter would do well to employ the use of a larger sample that could help to generalize findings to a larger population.

CONCLUSION

Coaching is a term often used to describe, "a collaborative, solution-focused, result-orientated and systematic process in which the coach facilitates the enhancement of life experience and goal attainment in the personal and/or professional life of… non-clinical clients" (Passmore, 2020). While it has existed within many other professions for some time (sports, mentoring, teaching, organizational leadership, psycho-social development, etc), it has only come to gain great popularity as a profession of its own, within the last two decades. This is with the promise of only exponentially increasing growth. In fact, it is reported that "the coaching industry is expected to grow at a compound annual growth rate of 7.6% (and) will be worth an estimated $27.5 billion by 2026" (Laker, 2023).

Despite legitimate concerns over lack of regulation and education/ training inconsistencies, coaching as a process, has been proven to be an effective relational approach to personal and professional growth. It has been proven to promote significant improvements in one or more health-related behaviors (Olsen & Nesbitt, 2010). It has also been associated with enhanced mental health, quality of life and goal attainment (Grant, 2003). One meta-analysis found that, when it came to executive coaching, there were "significant positive effects for some specific outcomes, such as self-efficacy, psychological capital, and resilience, indicating that executive coaching is effective in producing change even on dimensions considered relatively stable over time" (Nicolau et al., 2023).

There are beneficial aspects to coaching that promise to expand it as a professional competency and industrial lane. To do so effectively and responsibly, however, we must continue to ask difficult questions about what it is and what it is intended to do - and that we posit models and frameworks that can serve to organize the field's concepts, extend its paradigms and continue conversations around effective strategy and core responsibilities. This work is intended to contribute in just that way, asking professionals to consider a coaching model that at its core, works to improve the emotional intelligence of clients by targeting two important schema - how clients speak with themselves and how clients speak with others. The focus of this model is improving the emotional intelligence of clients' interpersonal and intrapersonal communications so as to advance them toward any goals they may set.

It is important to note that this is merely the beginning of a conversation. While theoretical frameworks are consulted and empirical studies are referenced, the posited models and accompanying case study are simply not sufficient to claim with any certainty that this model is exhaustive or validated. As we know, a case study is, "the detailed examination of a single example of a class of phenomena, a case study cannot provide reliable information about the broader class, but it may be useful in the preliminary stages of an investigation since it provides hypotheses, which may be tested systematically with a larger number of cases" (Flyvbjerg, 2011). Future researchers would do well to further explore, examine and test the models offered here with attention to their universal applications and nuances.

Still, this work does serve to contribute innovative perspectives to the ongoing research and conversation around coaching as a professional field, its modalities and its effectiveness. Utilizing a pracademic approach, analyzing a case study, employing the use of valid and reliable scales, this author puts forth an argument for coaching as an effective process in facilitating the improved emotionally intelligent interpersonal and intrapersonal communication of coaching clients. This work posits an effective coaching strategy in which the coach helps to facilitate a process of improving the emotional intelligence of client communication, both from an interpersonal and intrapersonal perspective.

REFERENCES

Bar-On, R. (1997). *BarOn Emotional Quotient Inventory: Technical Manual*. Multi-Health Systems.

Bhanji, F., Gottesman, R., de Grave, W., Steinert, Y., & Winer, L. R. (2012). The retrospective pre–post: A practical method to evaluate learning from an educational program. *Academic Emergency Medicine*, *19*(2), 189–194. DOI: 10.1111/j.1553-2712.2011.01270.x PMID: 22320369

Cai, D. A., & Fink, E. L. (2012). Communicate successfully by seeking balance. *Handbook of Principles of Organizational Behavior*, 425–444. DOI: 10.1002/9781119206422.ch23

Cooley, C. H. (1922). *Human nature and the social order*. Scribner.

Dush, D. M., Hirt, M. L., & Schroeder, H. E. (1983). Self-statement modification with adults: A meta-analysis. *Psychological Bulletin*, *94*(3), 408–422. DOI: 10.1037/0033-2909.94.3.408

English, S., Sabatine, J. M., & Brownell, P. (2019). *Professional coaching: Principles and practice*. Springer Publishing Company, LLC.

Erdös, T., de Haan, E., & Heusinkveld, S. (2020). Coaching: Client factors & contextual dynamics in the change process. *Coaching (Abingdon, UK)*, *14*(2), 162–183. DOI: 10.1080/17521882.2020.1791195

Ferrero, M., Vadillo, M. A., & León, S. P. (2021). A valid evaluation of the theory of multiple intelligences is not yet possible: Problems of methodological quality for intervention studies. *Intelligence*, *88*, 101566. DOI: 10.1016/j.intell.2021.101566

Flyvbjerg, B. (2011). Case Study. In N. K. Denzin & Y. S. Lincoln (Eds.), *The SAGE Handbook of Qualitative Research* (pp. 301–316). essay.

Gardner, H., & Hatch, T. (1989). Educational implications of the theory of multiple intelligences. [CrossRef] [Google Scholar]. *Educational Researcher*, *18*(8), 4–10. DOI: 10.3102/0013189X018008004

Goleman, D. (1998). *Working with Emotional Intelligence*. Bantam.

Goleman, D. (2007). *Emotional Intelligence* (10th ed.). Bantam Books.

Grant, A. M. (2003). The impact of life coaching on Goal Attainment, metacognition and Mental Health. *Social Behavior and Personality, 31*(3), 253–263. DOI: 10.2224/sbp.2003.31.3.253

Hardy, J. (2006). Speaking clearly: A critical review of the self-talk literature. *Psychology of Sport and Exercise*, *7*(1), 81–97. DOI: 10.1016/j.psychsport.2005.04.002

Higgins, E. T. (1989). Self-discrepancy theory: What patterns of self-beliefs cause people to suffer? *Advances in Experimental Social Psychology*, *22*, 93–136. DOI: 10.1016/S0065-2601(08)60306-8

Honeycutt, J. M., & Ford, S. G. (2001). Mental imagery and Intrapersonal Communication: A review of research on imagined interactions (IIS) and current developments. *Annals of the International Communication Association*, *25*(1), 315–345. DOI: 10.1080/23808985.2001.11679007

Jordan, M., & Livingstone, J. B. (2013, July). *Coaching vs psychotherapy in health and wellness: Overlap, dissimilarities, and the potential for collaboration*. Global advances in health and medicine. https://www.ncbi.nlm.nih.gov/pmc/articles/PMC3833547/

Kuhn, T. (2002). Negotiating boundaries between scholars and Practitioners. *Management Communication Quarterly*, *16*(1), 106–112. DOI: 10.1177/0893318902161008

Laker, B. (2023, July 21). *Every leader can benefit from coaching. Here's why...* Forbes. https://www.forbes.com/sites/benjaminlaker/2022/10/04/every-leader-can-benefit-from-coaching-heres-why/

Luft, J. A., Jeong, S., Idsardi, R., & Gardner, G. (2022). Literature Reviews, Theoretical Frameworks, and Conceptual Frameworks: An Introduction for New Biology Education Researchers. *CBE Life Sciences Education*, *21*(3), rm33. DOI: 10.1187/cbe.21-05-0134 PMID: 35759629

Mayer, J. D., Salovey, P., & Caruso, D. (2004). Emotional intelligence: Theory, findings and implications. *Psychological Inquiry*, *15*(3), 197–215. DOI: 10.1207/s15327965pli1503_02

Nicolau, A., Candel, O. S., Constantin, T., & Kleingeld, A. (2023). The effects of executive coaching on behaviors, attitudes, and personal characteristics: A meta-analysis of Randomized Control Trial Studies. *Frontiers in Psychology, 14*, 1089797. Advance online publication. DOI: 10.3389/fpsyg.2023.1089797 PMID: 37333584

Office for Human Research Protections (OHRP). (2023, May 31). *Exemptions (2018 requirements).* Retrieved from https://www.hhs.gov/ohrp/regulations-and-policy/regulations/45-cfr-46/common-rule-subpart-a-46104/index.html

Olsen, J. M., & Nesbitt, B. J. (2010). Health coaching to improve healthy lifestyle behaviors: An integrative review. *American Journal of Health Promotion, 25*(1), e1–e12. Advance online publication. DOI: 10.4278/ajhp.090313-LIT-101 PMID: 20809820

Passmore, J. (2020). Coaching defined and explored. In *The coaches' handbook: The complete practitioner guide for professional coaches* (First). essay, Routledge. DOI: 10.4324/9781003089889-2

Piaget, J. (1923). *Language and Thought of the Child.* Routledge.

Posner, P. L. (2009). The pracademic: An agenda for re-engaging practitioners and academics. *Public Budgeting &amp. Public Budgeting & Finance, 29*(1), 12–26. DOI: 10.1111/j.1540-5850.2009.00921.x

Raes, F., Pommier, E., Neff, K. D., & Van Gucht, D. (2011). Construction and factorial validation of a short form of the Self-Compassion Scale. *Clinical Psychology & Psychotherapy, 18*(3), 250–255. DOI: 10.1002/cpp.702 PMID: 21584907

Roberts, C. V. (1983, October 31). *The definition and delimitation of intrapersonal communication: A physiological perspective.* ERIC. https://eric.ed.gov/?id=ED240634

Schutte, N. S., Malouff, J. M., Hall, L. E., Haggerty, D. J., Cooper, J. T., Golden, C. J., & Dornheim, L. (1998). *Development and validation of a measure of emotional*

Schutz, W. (1958). *FIRO: A Three Dimensional Theory of Interpersonal Behaviour*. Rinehart.

Schutz, W. C. (1958). *FIRO: A three-dimensional theory of interpersonal behavior*. Rinehart.

Spitzberg, B. H. (1983). Communication competence as knowledge, skill, and Impression. *Communication Education, 32*(3), 323–329. DOI: 10.1080/03634528309378550

Spitzberg, B. H. (2015). Interpersonal Communication Competence and social skills. *The International Encyclopedia of Communication*, 1–8. DOI: 10.1002/9781405186407.wbieci078.pub2

Spitzberg, B. H., & Cupach, W. R. (1984). *Interpersonal communication competence* (Vol. 4). SAGE Publications, Incorporated.

Stone, J. (2016). Interpersonal Communication and Identity. In *Communication Competence: Creating Satisfying Interpersonal Relationships* (pp. 93–121). essay, University of Minnesota Libraries Publishing.

Stutts, L. (2022). Increasing Self-Compassion: Review of the Literature and Recommendations. *Journal of undergraduate neuroscience education: JUNE: a publication of FUN. Faculty for Undergraduate Neuroscience, 20*(2), A115–A119. DOI: 10.59390/WSZK3327 PMID: 38323068

Sullivan, H. S. (1953). *The Interpersonal Theory of Psychiatry*. W. W. Norton.

Vocate, D. R. (1994). *Intrapersonal communication: different voices, different minds*. Erlbaum.

Volpe, M. R., & Chandler, D. (1999). Resolving Conflicts in Institutions of Higher Education: Challenges for Pracademics. *CNCR-Hewlett Foundation Seed Grant White Papers*. 8. https://readingroom.law.gsu.edu/seedgrant/8

Vygotsky, L. S. (1930/1978). *Mind in Society: The development of higher psychological processes*. Harvard University Press.

Vygotsky, L. S. (1987). Thinking and Speech. In R. W. Rieber, & A. S. Carton (Eds.), The Collected Works of L. S. Vygotsky (Vol. 1), Problems of General Psychology (pp. 39-285). New York: Plenum Press. (Original Work Published 1934)

Zody, R. E. (1977). Academic/practitioner interchanges: Suggested guidelines. *Public Administration Review, 37*(5), 572. DOI: 10.2307/974719

Zourbanos, N., Hatzigeorgiadis, A., & Theodorakis, Y. (2007). A preliminary investigation of the relationship between athletes' self-talk and coaches' behaviour and statements. *International Journal of Sports Science &amp. International Journal of Sports Science & Coaching, 2*(1), 57–66. DOI: 10.1260/174795407780367195

Chapter 9
Catalyzing Aerospace Progress Machine Learning Strategies for Emerging Economics

Sreelekha Paul
https://orcid.org/0009-0007-5524-1919
Brainware University, India

Dipanjan Das
Brainware University, India

ABSTRACT

The advancement in machine learning technology is impacting the aircraft industry globally as it develops at a very high rate. That is why more and more emerging states come to realize that aerospace innovation is the key to economic diversification and technical self-sufficiency and the path to national development. Machine learning has played an important role in enhancing design, safety, and operation efficiency; classified learning, learning from experience learning, neural networks, and local learning are quite suitable here. Some of the methods of avoiding such negative impacts include localization of data generating projects as well as cloud computing options that have been developed to suit the socio-economic conditions of emerging countries and; ethical AI governance frameworks.

DOI: 10.4018/979-8-3693-7525-9.ch009

INTRODUCTION

Innovation as witnessed in the aerospace industry, has been the spear header of the national development Initiatives in scientific research, communication, transport and military. To assuage this, the industrialized nations have over the years maintained full control of the aerospace industry through better technological endowments, capital investment and environments for research. That said, the new hype and existence of digital disruption such as artificial intelligence, machine learning and data driven engineering as part of the disruption strategy have begun to make the playing field more leveled. New generations of emergent countries have more opportunities compared with the precedent ones to go beyond ordinary development models and start domestic innovations in aerospace industry and share in the international market. There are questions regarding the steps that emerging countries may take to alter their aerospace capabilities, contribute to improved economic stability, and help in the process of democratization of aerospace development with the use of digital technology.

Background

This is particularly evidenced by the fact that as more and more developing economies are now emerging to come up with and endorse state of the art utensils for defining their strategic relevance, the aerospace landscape internationally is gradually evolving. These countries might move from historical limitations and adopt technologically advanced fields such as satellite launch, usage of drones and space discovery due to digitization, infrastructure development and education (OECD, 2020). Aerospace innovation, therefore, is a process by which developing nations achieve diversification of their economy, a guarantor of their security and a way to become a player on the world stage and to enhance their technological progress. ML, a component of the Artificial Intelligence (AI) that designs systems to learn from experience, may change the aircraft industry. (Jordan et al., 2015 and Mitchell, 2015) claim that ML enhances aircraft operational performance, optimizes design processes, offers reliable maintenance and enhances the operational safety. Machine learning can help emerging economies enhance their aeronautical capacities and advance the innovation cycle occurrences.

MACHINE LEARNING APPLICATIONS IN AEROSPACE

Since ML helps to perform operations safer, faster and more intelligently, it has become one of the significant driving factors in aerospace systems change. They range from design of materials and structures, structural health, monitoring of facilities, intelligent drones and satellites, optimization of air traffic control in real-time, and prediction of repair of aircraft engines, mostly aircrafts (Shukla & Mishra, 2020). Federated learning also helps the aerospace organization to conduct collaboration to form big data without sharing the data. Each of them contributes to improving efficiency, expanding capabilities and accelerating product development with regard to military and civil aerospace fields.

Key Machine Learning Techniques in Aerospace

In aerospace engineering there are now seen impacts of machine learning techniques as these give enhanced foreseeable capabilities, enhanced and effective functions, and separation of autonomous systems. In the case of predictive maintenance using analytical models such as wear and tear of the engine and likely failure of some components using input from sensors, labeled data sets are used in supervised learning (Zhao et al., 2019). For example, Rolls-Royce's "Intelligent Engine" program employs supervised machine learning algorithms to monitor the status of engines onboard different aircrafts in a real-time manner while this approach helps to reduce both time, which is spent on maintenance, and direct costs associated with it.

In the case of the OFP, acquisition learning based on rewards, which is the basis of RL, is vital. As mentioned by (Kochenderfer et al., 2019 and Wheeler 2019), RL is capable of changing mission objectives, obstacles, and fuel requirements of the flight paths in real time.

It was found out that deep learning and particularly multilayered neural networks are perfect for analyzing satellite pictures. For example, (Kussul et al., 2017) noted that CNNs have boosted the capability of recognizing objects in contexts such as identifying patterns of the use of land and monitoring modifications to the environment based on data from orbiting systems.

Furthermore, the use of federated learning is gradually getting important in the aerospace industry since it enables collaborative model training across distributed sources while protecting data privacy. In the opinion of Li et al. (2020), this is more relevant for transmitting satellite data across organizations as well as the use of sensitive applications in the defense aerospace sector.

Major Areas of ML Integration

As an effective and powerful tool, machine learning (ML) is more and more applied into technique and a significant part of aerospace engineering since it can enhance efficiency, safety and performance. Among the real-life applications of machine learning to enhance flight control, autopilot systems, fault detection, and control adaptive flight systems and avionics, the algorithms predict the occurrence of domains and correct them on their own (Zhang et al., 2019). For instance, Boeing's Airplane Health Management System uses ML for inspecting the data of the airplane's sensors and recommending actions to be taken.

In their opinion, (Bendsøe et al., 2013 and Sigmund, 2013) stated that engineers are able to cut down the development cycles while enhancing the fuel economy in the designs concerning aircraft and other structures and materials by introducing more effective generative machine learning models useful in simulating complex structures in aerodynamics and light and thin laminates. Although the stress contours are non-linear in new-age aerospace materials, deep learning can be applied to it effectively.

ML is also useful in air traffic management systems most of the time when it comes to predicting flight route, traffic, and weather delays. Decision-support technologies such as machine learning, leverage traffic and reduce traffic congestion (Sun et al., 2021).

Risk management is also made possible through the use of past performance to predict the times at which certain parts are likely to fail, plan for their maintenance and in turn anticipate similar problems in logistics. This is best illustrated by Airbus's Skywise, where data of the entire fleet is analyzed to offer suggestions that reduce maintenance cost and time.

STATES OF AEROSPACE AND ML ADOPTION IN DEVELOPING ECONOMICS

Through targeted investments and a rising interest in machine learning (ML) applications, developing economies are adopting aerospace innovation more and more. Using machine learning (ML), nations such as Brazil, South Africa, and India are developing UAV systems, satellite technology, and predictive analytics to close technical gaps. The lack of trained personnel, infrastructure, and data availability, however, means that growth is still unequal (OECD, 2020). In spite of these obstacles, scholarly research, international alliances, and public-private partnerships are promoting the use of machine learning (ML), allowing these countries to compete globally and develop their own aerospace solutions.

Current Aerospace Capabilities

It is worth underlining the fact that various developing nations are improving their aeronautical capacities and capabilities through classifications, both in the private sector and by means of government-proposed programs. In Chandrayaan and Mangalyaan missions, the aerospace industry of India has outlined itself as a capable contender at the international level with the help of Indian Space Research Organization (ISRO). The budding space industry of India comprises Sky root Aerospace and fulfills the country's aspirations of having new inventions and launch vehicles for satellite deployment (Ravindran et al.,, 2022).

Today, Brazil conducts space activities through the Brazilian Space Agency that cooperates with other countries on satellite and propulsion projects; also, through one of the world largest regional aircraft manufacturers 'Embraer' the country remains an active participant of the Aerospace industry (OECD, 2020).

Through SANSA, which is a participant in the SKA, a global radio telescope project, the country has enhanced its capacity in satellite engineering and earth observation (SANSA, 2021). Nigeria is still relatively new, but through the NASRDA, it has sent many satellites largely with communication and agricultural observation intentions (Adewale et al., 2020 & Akinwale, 2020). LAPAN of Indonesia (now BRIN) has contributed to space research while the country's commercial side began broaching UAVs and atmospheric monitoring systems. These countries' capabilities are diverse and growing based on the increasing demand for artificial intelligence (AI) and machine learning for aerospace.

Challenges Faced

Hiring of ML as an ingredient in aerospace innovation is a perennial issue for emergent countries as much as they aspire. Lack of resources and funding are the major issues stopping the development of complex simulating labs, test beds and data centers. Most of the poor nations spend less than 1% of their GDP on knowledge capital and this slows down long-term aerospace advancement, as pointed out by the World Bank (2021).

There is the deficiency of talent or even what could be referred to as the brain drain which is another weakness. There are not enough engineers trained in ML and aerospace science in the country; while countries like Nigeria and India produce high caliber engineers, many of them emigrate to developed countries in search of employment opportunities for carrying out research (UNESCO, 2021).

Training of ML models requires large datasets, and high computation power which itself due to several reasons including limited access to big data and GPU is not feasible in their training. For example, projects in African sub-region most government space programs rely on foreign partners to get satellite data and analytically platforms (Adewale et al., 2020 and Akinwale, 2020). Together, these concerns cumulatively hamper the implementation of ML for facilitating scalable innovation in emerging nations and hence restrict the aerospace industry.

Adaptive Gradient Method

The following structures observed in emerging economies are likely to enhance the application of ML in aerospace innovation: Among these the most important are their youthful and technically inclined populations. For instance, over 60 percent of Africans are under 25 years, and they are receiving enhanced access to education especially STEM and getting in contact with digital technology thus creating a good base for future prospects of AI and aerospace (UNDP, 2020).

On the same note, greenfield innovation prospects are attained here whereby these nations get a chance to move to other phases of aerospace development apart from being bounded by old structures. Forced learning in IT and data analysis for private sector firms in India, explains how it is possible to adopt state of the art technology in the areas of UAV navigation, satellite analytics and smart manufacturing' (Ravindran et al., 2022).

It is also concerning the International collaborations, and effort made by other countries following Diaspora networks. Such schemes include GMES & Africa of the African Union, cooperation on Space Technology and Innovation between India and JAXA and between NASA and India. The diaspora professionals also contribute indirectly through participating in investments in domestic aerospace ventures, technology transfer and sponsoring of other firms and start-ups (OECD, 2020). All these factors place rising economies in a capacity to flexibly and creatively participate in the global aerospace scene.

CASE STUDIES OF MACHINE LEARNING-DRIVEN AEROSPACE INNOVATION

Based on some given cases, it is possible to examine how the application of ML influences the increase in aerospace innovations in emerging economies. To reduce the design time and cost, an organization situated in India, uses machine learning approaches to build the model of multiple propulsion systems for the small satellite launch vehicles. For the enhancement of the dependability of aircraft in Brazil's Embraer, ML is used to detect the defects and analyze the real-time flight data. Machine learning is used by satellites in South Africa for improving its earth observation in agricultural planning and climate changes. As mentioned by OECD (2020) & Ravindran et al., 2022), the emerging economy can leverage AI to compete globally and address the local socio-economic, and environmental challenges. These activities are some of the many that compose an emerging trend of a more strategic buy and use of goods from ML.

India: ISRO and Satellite Data Analytics

DL and ML are being incorporated in ISRO, which is the governing body of the Indian space program, for betterment in the proficiency of its missions and for processing the received satellite data. One of the key application areas of deep learning is in positioning surveillance for agriculture as well as categorization of the land. CNNs have been adopted by ISRO along with academic institutions to classify crop types, irrigation facilities, and soil moisture from remote-sensing data available from IRS and Cartosat series. This is well desired for the Indian agricultural sector (Dadhwal et al., 2020 & Parihar, 2020). Examples of such models include the FASAL – an application that stands for Forecasting Agricultural Output using Space, Agro-meteor and Land observation. In minimizing risk and optimizing the possibility of fuel consumption, they apply ML to corridors by simulating the propulsion dynamics as well as environmental and payload parameters (Ravindran et al., 2022). All of these are forms of System of Systems (SoS) where aeronautical systems are increasingly becoming data-driven and self-sufficient. India is creating a synthesis of ISRO technological expertise and the private sector's agility in the application of ML daily in the commercial domain and space technology applications.

Brazil: Aircraft Maintenance and Inspection Automation

Vision-based machine learning (ML) systems for automated inspection and structural health monitoring of airplanes are being used in Brazil and it is leading the way. Brazil is the country where one of the world's largest aircraft manufacturers is located – Embraer. Through detecting surface cracks, corrosion, and wear on airframes by way of computer vision and deep learning, these technologies increase the security of flights, shortening the time and cost of maintenance in the process. For identifying microstructural defects that may be missed out by human inspectors, for instance, Embraer has developed ML enabled inspection drones that are equipped with LiDAR sensors and high-resolution cameras. These drones scan the surface of the aircraft automatically (Silva et al., 2020). While using datasets collected during several decades of aircraft operations, these vision-based systems also rely on convolutional neural networks (CNNs) for predicting the structural wear and recognizing various types of damage (Nascimento et al., 2021). Through this practice, the downtime of aircrafts is reduced and life of the critical components is enhanced. With these advances, Embraer has managed to secure its place as a leader in the world of aviation technology due to the fact that these development trends reflect Brazil's strategic goal of incorporating machine learning (ML) into aerospace production and the global trend to introduce digital twins and data-driven maintenance systems into manufacturing.

Nigeria: Startups and UAV Development

Nigeria's booming aerospace industry is being propelled on by startups that are building unmanned aerial vehicles (UAVs) for security, agriculture, and logistics. Arone and QuadLoop are some of the businesses that are implementing machine learning (ML) algorithms to assist UAV systems to make real time decisions and track paths whilst operating autonomously. The mission of such companies is to create cheap energy-saving drones ideally suitable for the rugged and constantly changing face of Africa (Adewale et al., 2021). For instance, the drone navigation done by the computer vision of Arone's drones embraces machine learning (ML) to enable precise perception of obstacles and flight avoidance in urban areas. These systems, doing visual input assessment, use convolutional neural networks (CNNs), increasing situational awareness and providing the increase of the chance of collisions. (Okafor et al., 2020 & Nwosu, 2020) For the last mile delivery in the remote areas, QuadLoop does integrate ML-powered route analysis, which optimizes the flight time while using less power. To enhance efficiency and safety Nigerian UAV developers are also researching the use of reinforcement learning (RL) for dynamic mission planning for which the drones learn to change their flying route in real-time, in the light of environmental data. These developments are manifestations of the Nigerian drive for becoming a leader in UAV technology where applications in its rapidly growing economy will include precision agriculture and disaster response.

ENABLERS FOR SUCCESSFUL ML-AEROSPACE INTEGRATION

There are certain essential enablers that need to be implemented in order to be able to integrate machine learning (ML) successfully into aerospace. Data infrastructure has to be quite advanced in order to process and analyse the vast amounts of aircraft data in real time. The process of ML model training is expedited by investments in secure cloud platforms and high-performance computing (HPC) abilities. It is through cross-sector partnership between the government, business and academia that knowledge transfer and innovation are encouraged. Besides, competent human resources are required; therefore, the need to train in specialized courses in machine learning and aeronautical engineering arises (OECD, 2020). Policies that facilitate the cooperation of the world and open data access also enhance capacities and thus emerging economies can effectively benefit from the global knowledge.

Policy and Regulatory Support

Rapid introduction of the machine learning (ML) on aircrafts, particularly in emerging economies, demands effective legislative and regulatory benchmarks. Open data efforts, one of the key prerequisites, make extensive datasets demanded for precise machine learning models available. For instance, the Open Access Hub of the European Union's Copernicus has been a reference point in other places and has been useful in advancing the cause of environmental monitoring and satellite analytics (European Commission, 2020). Such laws promoting the sharing of data for urban planning, agriculture and disaster management from satellites may be beneficial for emerging economies.

A framework of aerospace AI should also be developed for standardization of developing, testing, and implementing machine learning systems. In order to preserve public confidence as well as global interoperability, these standards need to address safety, cybersecurity, and data privacy (ICAO, 2021). Techniques for integrating machine learning (ML) into critical flight systems have been issued by the European Aviation Safety Agency (EASA) and U.S. Federal Aviation Administration (FAA) in the form of guidelines (EASA, 2022). Similar structures could be used by emergent nations to achieve ethical and secure use of AI as they build an internationally competitive aerospace industry.

Talent Development

Major investment is required in the aerospace industry in STEM education and reskilling ventures in order to integrate machine learning. In India, the Atal Innovation Mission and Science Without Borders, in Brazil, are two of the initiatives that have opened up technical education and propelled the young engineers towards AI and aeronautic technology. The real application in areas such as satellite analytics, autonomous flying, and predictive maintenance of machine learning (ML) is advanced through the cooperation of universities and industries including research collaboration between academic institutions and aerospace corporations. By using these initiatives, the human capital needed for long term aeronautical invention is reinforced.

Infrastructure and Digital Ecosystem

A robust digital infrastructure is necessary for the incorporation of machine learning (ML) in the aircraft industry due to its processing strength needed to support the huge amounts of satellite images as well as real-time flight data. Part of the instrumental components is cloud computing that offers a scalable and low-cost access to high-performance computing (HPC) for complex machine learning model training. For example, there are the on-demand GPU clusters provided by the platforms such as Microsoft Azure and Amazon Web Services (AWS) to the aerospace companies where the upfront costs are reduced, enhanced by the wave of innovation (AWS, 2022). Cloud services for rapid design cycles are relied on by firms such as Skyroot Aerospace while, in India, ISRO has utilized cloud-based HPC for analyzing satellite data (Ravindran et al., 2022).

Affordable access to supercomputing resources is just as important. Nations such as Brazil have invested in national HPC facilities such as Santos Dumont Supercomputer that aid in the study of the fluid dynamics and simulation aerospace (Nascimento et al, 2021). Robust cybersecurity practices are essential in maintaining operational propriety and confident crusts in light of connected aircrafts and digital twins (ICAO, 2021). This Infrastructure provides real time analytics and data driven decision making which helps as a foundation for modern aerospace innovation.

ETHICAL, SOCIAL AND SOCIAL CONSIDERATION

There are important ethical, legal, and sociological concerns that arise when machine learning (ML) is used in aerospace industries. Legal rules should be put in place to ensure that people are accountable for ML systems, and ML models should remain explainable and open to maintain public trust. In addition, as long as unequal access to data and computing powers exists, the use of machine learning in aerospace could increase digital inequality (OECD, 2020). To ensure aerospace AI remains sustainable and responsible, we need to strike a balance between innovation and keeping an eye on ethics.

Data Privacy and Sovereignty

Good data management is necessary for both the security of the country and the productive use of resources in the aviation sector. Such data includes what is recorded by mission control, the signals sent during a flight, and satellite imagery, all of which can be a high security threat if someone gains access. Having unauthorized access to satellite positioning could affect essential services, while letting go of real-time information about an aircraft might endanger the country's airspace (ICAO). With aerospace organizations relying more on international cloud systems for their data, keeping data within one nation's borders is becoming a bigger concern. As a precaution under data privacy laws in India, the ISRO limits the sharing of sensitive satellite data with foreign servers by handling it by themselves (Ravindran et al., 2022). In order to maintain control over their citizens' information, the European Union installed the General Data Protection Regulation (GDPR), and this could act as a guide for other countries (European Commission, 2020). To deal with these issues, aerospace companies rely on multi-factor authentication, complete encryption of data, and avoid trusting the entire network. These steps give the public confidence in AI planes, make sure they comply with all regulations, and strengthen how secure they are.

Advanced Techniques

Pursuing successful hyper-parameter settings, Bayesian optimization utilizes probabilistic models to predict prior assessments towards the probable performance within certain territories (Snoek et al. 2012). Differential evolution and other evolutionary algorithms produce candidate solutions over multiple iterations, learning Differential evolution and other evolutionary algorithms are well suited to high-dimensional and non-linear problems given their ability to in learning complicated and even multimodal search spaces, differential evolution outperforms almost all other state-of-the-art stochastic global optimization techniques (Real et al. 2017).

FUTURE TRENDS AND OPPORTUNITIES

It is thanks to Edge AI that small satellites and drones are enabled to handle real-time decisions, without relying on ground computers. Thanks to lower latency, the method increases operational stability. As an example, NASA's Mars Helicopter Ingenuity depends on on-board AI to sense hazards and find a way through environments where communication is scarce (NASA, 2021). For use in both business and military settings, Skydio drones use edge AI to stay clear of obstacles in difficult areas

(Skydio, 2022). Reconnaissance on the front line and missions into space have become possible thanks to them.

Through algorithms that maximize the weight, strength, and efficiency in designs, generative AI is revolutionizing the aircraft design field. As an example, Airbus designs airplane pieces based on nature and using generative design to make them lighter and more fuel efficient (Airbus, 2020). They study tons of variations to propose new and better layouts compared to regular designs based on performance. Not only does it help aerospace companies reach their goals for sustainable manufacturing, but it also increases the speed of creating prototypes and reduces the amount of wasted materials.

The industry is excited about QML because it can solve tough aerodynamic problems more quickly than previous approaches. The use of QML in aircraft design can allow for faster cycles due to its ability to model fluid dynamics, structural stress, and heat easily (Preskill, 2018). Google and D-Wave are currently researching applications of QML for CFD, and it is possible that this could help improve how turbulence is handled or flying at high speeds à la hypersonic planes (D-Wave, 2022). With its ability to speed up invention, this technology could transform how aerospace R&D is carried out.

Real progress in using AI in aerospace requires collaboration between several nations. South-South cooperation is becoming more important since it focuses on partnerships among developing countries. One instance of South Africa and India teaming up in space is the South African CubeSat ZACube-2 (SANSA, 2021). They help make the aerospace industry more open by encouraging this collaboration.

CONCLUSION

Overall, machine learning (ML) can significantly improve the aerospace industries of emerging nations by helping them gain an edge, promote progress, and minimize workflow problems. Following the key results, these countries can rely on machine learning (ML) for air traffic management, preventing equipment failures, pilotless flight, and studying satellite data consistent with worldwide aerospace trends. Still, the country needs effective technology, talented people, and the right policies to realize these opportunities. Assuring that AI systems are open, safe, and follow good values means national ML plans need to focus a lot on keeping data protected, keeping computer networks secure, and growing the number of people who know how to work with AI. OECD, 2020). In order to sustain innovation ecosystems, policymakers should support the building of fast computers, help schools and businesses work together, and offer help to new businesses getting started. In conclusion, we need to help growing economies play a bigger role in aerospace technology, so progress can be made in a way that doesn't harm the environment and so that countries can build their own technology. This plan aims to ensure community-wide creation of blueprints for the field, which will give the country an advantage in strategic interests and help its economy grow.

REFERENCES

Adewale, A. A., & Akinwale, O. O. (2020). Assessment of Nigeria's space programme: Challenges and prospects. *Space Policy*, *51*, 101335. DOI: 10.1016/j.spacepol.2019.101335

Adewale, A. A., & Akinwale, O. O. (2020). Assessment of Nigeria's space programme: Challenges and prospects. *Space Policy*, *51*, 101335. DOI: 10.1016/j.spacepol.2019.101335

Adewale, A. A., Okafor, E. O., & Nwosu, C. N. (2021). Innovations in drone technology: Impacts and prospects for developing economies. *African Journal of Science, Technology, Innovation and Development*, *13*(4), 551–560. DOI: 10.1080/20421338.2021.1889363

Airbus. (2020). *Generative design: The future of aerospace engineering*. https://www.airbus.com

AWS. (2022). *Cloud computing for aerospace and satellite solutions*. https://aws.amazon.com/aerospace

Bendsøe, M. P., & Sigmund, O. (2013). *Topology optimization: Theory, methods, and applications* (2nd ed.). Springer.

D-Wave. (2022). *Quantum computing for aerospace*. https://www.dwavesys.com

Dadhwal, V. K., & Parihar, J. S. (2020). Space-based crop monitoring in India: The FASAL experience. *Current Science*, *119*(6), 940–949.

European Commission. (2020). *Copernicus Open Access Hub*. https://scihub.copernicus.eu

European Commission. (2020). *General Data Protection Regulation (GDPR)*. https://ec.europa.eu

European Union Aviation Safety Agency (EASA). (2022). *AI roadmap: A human-centric approach to AI in aviation*. https://www.easa.europa.eu

ICAO. (2021). *Cybersecurity in civil aviation: A global perspective*. International Civil Aviation Organization.

International Civil Aviation Organization (ICAO). (2021). *Manual on machine learning for aviation safety*. ICAO Publishing.

International Civil Aviation Organization (ICAO). (2021). *Manual on machine learning for aviation safety*. ICAO Publishing.

International Civil Aviation Organization (ICAO). (2021). *Manual on machine learning for aviation safety*. ICAO Publishing.

Jordan, M. I., & Mitchell, T. M. (2015). Machine learning: Trends, perspectives, and prospects. *Science*, *349*(6245), 255–260. DOI: 10.1126/science.aaa8415 PMID: 26185243

Kochenderfer, M. J., & Wheeler, T. A. (2019). *Algorithms for decision making*. MIT Press.

Ministry of Education. (2021). *Annual report on technical education and skill development in India*. Government of India.

NASA. (2021). *Ingenuity Mars Helicopter – Technology Demonstration*. https://mars.nasa.gov

Nascimento, D. L., Silva, R. G., & Martins, M. F. (2021). Machine learning applications in aircraft structural health monitoring. *Journal of Aerospace Technology and Management*, *13*(1), e2021–e1213. DOI: 10.5028/jatm.c2021-1213

OECD. (2020). *The future of the space economy: Trends, challenges and growth opportunities*. OECD Publishing., DOI: 10.1787/c5996201-

OECD. (2020). *The future of the space economy: Trends, challenges and growth opportunities*. OECD Publishing., DOI: 10.1787/c5996201-

Okafor, E. O., & Nwosu, C. N. (2020). Machine learning in autonomous UAV navigation: Trends and future directions. *Journal of Intelligent & Robotic Systems*, *99*(3), 421–435. DOI: 10.1007/s10846-020-01215-8

Organisation for Economic Co-operation and Development. (2020). *The future of the space economy: Trends, challenges and growth opportunities*. OECD Publishing., DOI: 10.1787/c5996201-

Preskill, J. (2018). Quantum computing in the NISQ era and beyond. *Quantum : the Open Journal for Quantum Science, 2*, 79. DOI: 10.22331/q-2018-08-06-79

Ravindran, S. (2022). The rise of private space in India. *Nature India, 18*, 324–325. DOI: 10.1038/nindia.2022.140

Silva, R. G., Sousa, A. C. M., & Lima, M. R. (2020). Automated drone inspections in aerospace: Challenges and future directions. *Aerospace Science and Technology, 106*, 106078. DOI: 10.1016/j.ast.2020.106078

Skydio. (2022). *Skydio autonomous flight systems*. https://www.skydio.com

South African National Space Agency (SANSA). (2021). Annual report 2020/2021. https://www.sansa.org.za

South African National Space Agency (SANSA). (2021). *South African satellite missions*. https://www.sansa.org.za

Sun, D., Zheng, S., Li, Y., & Jia, M. (2021). A machine learning approach to air traffic flow management. *Transportation Research Part C, Emerging Technologies, 129*, 103254. DOI: 10.1016/j.trc.2021.103254

UNESCO. (2021). Science report: The race against time for smarter development. UNESCO Publishing. https://unesdoc.unesco.org/ark:/48223/pf0000377257

United Nations Development Programme (UNDP). (2020). *Human development report 2020: The next frontier—Human development and the Anthropocene*. UNDP. https://hdr.undp.org

World Bank. (2021). World development indicators. https://data.worldbank.org

Zhang, Y., Yang, G., & Zhang, Y. (2019). Intelligent fault diagnosis of aerospace systems using machine learning: Review and perspectives. *Chinese Journal of Aeronautics, 32*(1), 213–236. DOI: 10.1016/j.cja.2018.09.012

Zhao, R., Yan, R., Chen, Z., Mao, K., Wang, P., & Gao, R. X. (2019). Deep learning and its applications to machine health monitoring: A survey. *Mechanical Systems and Signal Processing, 115*, 213–237. DOI: 10.1016/j.ymssp.2018.05.050

Compilation of References

Aburaya, A., Selamat, H., & Muslim, M. T. (2024). Review of vision-based reinforcement learning for drone navigation. *International Journal of Intelligent Robotics and Applications*, *8*(4), 1–19. DOI: 10.1007/s41315-024-00356-9

Action, J., Morehead, K., Hiromoto, T., & Gajus, B. (2024). Application of Generative Design to Thin-Walled Aerostructure. In AIAA SCITECH 2024 Forum (p. 0362). DOI: 10.2514/6.2024-0362

Acuña, J. (2021). Evaluación limnológica y microbiológica de la Laguna del Quilotoa, provincia de Cotopaxi – Ecuador. *UNIVERSIDAD CENTRAL DEL ECUADOR*, *14*, 99–100.

Adewale, A. A., & Akinwale, O. O. (2020). Assessment of Nigeria's space programme: Challenges and prospects. *Space Policy*, *51*, 101335. DOI: 10.1016/j.spacepol.2019.101335

Adewale, A. A., Okafor, E. O., & Nwosu, C. N. (2021). Innovations in drone technology: Impacts and prospects for developing economies. *African Journal of Science, Technology, Innovation and Development*, *13*(4), 551–560. DOI: 10.1080/20421338.2021.1889363

Adimurthy, V., Balu, R., Priyadarshi, P., Ramanan, R. V., & Ravikumar, C. (2005). Multidisciplinary Optimization Aerospace Design: The Emerging Technology For Complex Systems. *Journal of Aerospace Sciences and Technologies*, •••, 79–92.

Aerospace Quality Trends in AI-Driven Manufacturing. (2023). *PwC Industry Report.*

Agarwal, A., Ajith, A., Wen, C., ... (2023). Robotic defect inspection with visual and tactile perception for large-scale components. *arXiv preprint arXiv:2309.04590.* DOI: 10.1109/IROS55552.2023.10341590

AI in Aerospace: Opportunities and Challenges. (2023). *McKinsey & Company Report.*

AI-Augmented NDT in Aerospace: A Review. (2023). *Journal of Non-Destructive Testing,45*(2), 89-105.

AI-Based Predictive Maintenance for Aircraft Components. (2023).. . *Reliability Engineering & System Safety*, *95*(4), 460–478.

Airbus. (2020). *Generative design: The future of aerospace engineering.* https://www.airbus.com

Amune, A., Shahari, S., Kasurde, S., Nimale, S., Surdas, S., & Tayde, S. (2024, April). Exploring Predictive Maintenance and Signal Processing Techniques for Automotive Health Monitoring. In *2024 International Conference on Expert Clouds and Applications (ICOECA)* (pp. 541-547). IEEE. DOI: 10.1109/ICOECA62351.2024.00100

Anderson, M., & Brown, C. (2023). Integration of computer vision systems in aircraft assembly verification. *Aviation Safety Journal*, *27*(4), 678–689.

Anjana, P. K., Abhilash, P. V., Bisariya, S., & Sutrakar, V. K. (2024). Inverse Approach for Metasurface Based Radar Absorbing Structure Design for Aerospace Applications Using Machine Learning Techniques (No. 2024-26-0480). SAE Technical Paper.

Arafat, M. Y., Alam, M. M., & Moh, S. (2023). Vision-based navigation techniques for unmanned aerial vehicles: Review and challenges. *Drones (Basel)*, *7*(2), 89. DOI: 10.3390/drones7020089

Arista, R., Mas, F., Morales-Palma, D., Oliva, M., & Vallellano, C. (2022). The role of Industrial Resources in Reconfigurable Aerospace Production Systems: A Preliminary Literature Review. In *IFAC-PapersOnLine* (Vol. 55, pp. 2719–2724). Elsevier B.V., DOI: 10.1016/j.ifacol.2022.10.124

Aust, T., & Pons, D. (2022). Automated defect detection on engine blades: A comparative study. *Journal of Aerospace Engineering*, *36*(4), 1–15.

AWS. (2022). *Cloud computing for aerospace and satellite solutions.* https://aws.amazon.com/aerospace

Badrloo, S., Varshosaz, M., Pirasteh, S., & Li, J. (2022). Image-based obstacle detection methods for the safe navigation of unmanned vehicles: A review. *Remote Sensing (Basel)*, *14*(15), 3824. DOI: 10.3390/rs14153824

Bakirci, M., & Bayraktar, I. (2025). *AI-Driven Micro Defect Detection for Aerospace*. Neural Concept Research Brief.

Balani, R. S., Patel, D., Barthwal, S., & Arun, J. (2021). Fabrication of Air Conditioning System Using the Engine Exhaust Gas. *Advances in Interdisciplinary Research in Engineering and Business Management*, 369-378.

Barkalov, K., Lebedev, I., Usova, M., Romanova, D., Ryazanov, D., & Strijhak, S. (2022). Optimization of turbulence model parameters using the global search method combined with machine learning. *Mathematics*, *10*(15), 2708. DOI: 10.3390/math10152708

Barlow, C., & Peterson, R. (2022). *Artificial Intelligence in Aerospace: Applications and Challenges*. Springer.

Bar-On, R. (1997). *BarOn Emotional Quotient Inventory: Technical Manual*. Multi-Health Systems.

Beetham, S., & Capecelatro, J. (2020). Formulating turbulence closures using sparse regression with embedded form invariance. *Physical Review Fluids*, *5*(8), 084611. DOI: 10.1103/PhysRevFluids.5.084611

Beltrán-González, M., Smith, J., & Lee, K. (2020). Machine learning-based visual inspection system for aerospace component defect detection. *International Journal of Advanced Manufacturing Technology*, *110*(9-12), 2801–2813.

Bendsøe, M. P., & Sigmund, O. (2013). *Topology optimization: Theory, methods, and applications* (2nd ed.). Springer.

Bertoni, A., Hallstedt, S. I., Dasari, S. K., & Andersson, P. (2020). Integration of value and sustainability assessment in design space exploration by machine learning: An aerospace application. *Design Science*, *6*, e2. Advance online publication. DOI: 10.1017/dsj.2019.29

Be co, D., Pârvu, P. V., & Ciudin, S. (2024). Autonomous navigation for the Martian Inspection Drone. *Acta Astronautica*, *216*, 55–63. DOI: 10.1016/j.actaastro.2023.12.052

Bhanji, F., Gottesman, R., de Grave, W., Steinert, Y., & Winer, L. R. (2012). The retrospective pre–post: A practical method to evaluate learning from an educational program. *Academic Emergency Medicine*, *19*(2), 189–194. DOI: 10.1111/j.1553-2712.2011.01270.x PMID: 22320369

Bidar, O., He, P., Anderson, S., & Qin, N. (2024). Aerodynamic shape optimisation using a machine learning-augmented turbulence model. In *AIAA SCITECH 2024 Forum* (p. 1231). DOI: 10.2514/6.2024-1231

Bijjahalli, S., Sabatini, R., & Gardi, A. (2020). Advances in intelligent and autonomous navigation systems for small UAS. *Progress in Aerospace Sciences*, *115*, 100617. DOI: 10.1016/j.paerosci.2020.100617

Bishop, C. M. (2006). *Pattern recognition and machine learning*. Springer.

Boopathy, G., Gurusami, K., Chinnapandian, M., & Vijayakumar, K. R. (2022). Optimization of Process Parameters for Injection Moulding of Nylon6/SiC and Nylon6/B_4C Polymer Matrix Composites. *Fluid Dynamics & Materials Processing*, *18*(2), 223–232. DOI: 10.32604/fdmp.2022.018225

Boopathy, G., Palani, S., Ganesan, B., & Kumaar, K. (2024). Advancements in Design, Fabrication, and Applications of Functionally Graded Composite Materials. In *Engineering Materials for Efficient Energy Storage and Conversion* (pp. 25–50). IGI Global. DOI: 10.4018/979-8-3693-2798-2.ch002

Borwankar, P., Kapania, R. K., Inoyama, D., & Stoumbos, T. (2024). Multidisciplinary Design Analysis and Optimization of Space Vehicle Structures. In *AIAA SCITECH 2024 Forum* (p. 2898). DOI: 10.2514/6.2024-2898

Bounds, C. P., Uddin, M., & Desai, S. (2023). Tuning of Turbulence Model Closure Coefficients Using an Explainability Based Machine Learning Algorithm (No. 2023-01-0562). SAE Technical Paper.

Boy, G. A., & Morel, C. (2022). The machine as a partner: Human-machine teaming design using the PRODEC method. *Work (Reading, Mass.)*, *73*(s1), S15–S30. DOI: 10.3233/WOR-220268 PMID: 36214030

Brenner, M. P., Eldredge, J. D., & Freund, J. B. (2019). Perspective on machine learning for advancing fluid mechanics. *Physical Review Fluids*, *4*(10), 100501. DOI: 10.1103/PhysRevFluids.4.100501

Brunton, S. L., & Kutz, J. N. (2022). *Data-driven science and engineering: Machine learning, dynamical systems, and control*. Cambridge University Press. DOI: 10.1017/9781009089517

Brunton, S. L., Nathan Kutz, J., Manohar, K., Aravkin, A. Y., Morgansen, K., Klemisch, J., & McDonald, D. (2021). Data-driven aerospace engineering: Reframing the industry with machine learning. *AIAA Journal*, *59*(8), 2820–2847. DOI: 10.2514/1.J060131

Butler, G. A., Gu, D., Humfeld, K. D., ... (2021). A machine learning framework for real-time inverse modeling and multi-objective process optimization. *arXiv preprint arXiv:2104.11342*.

Butler, G. A., Gu, D., Humfeld, K. D., Nelson, K., & Zobeiry, N. (2021). A machine learning framework for real-time inverse modeling and multi-objective process optimization of composites for active manufacturing control. *arXiv preprint arXiv:2104.11342*.

Cai, D. A., & Fink, E. L. (2012). Communicate successfully by seeking balance. *Handbook of Principles of Organizational Behavior*, 425–444. DOI: 10.1002/9781119206422.ch23

Caliari, T., & Ferreira, M. J. B. (2023). The historical evolution of the Brazilian aeronautical sector: A combined approach based on mission-oriented innovation policy (MOIP) and sectoral innovation system (SIS). *Economics of Innovation and New Technology, 32*(5), 682–699. DOI: 10.1080/10438599.2021.2011258

Caliari, T., Ribeiro, L., Pietrobelli, C., & Vezzani, A. (2023). Global value chains and sectoral innovation systems: An analysis of the aerospace industry. *Structural Change and Economic Dynamics, 65*, 36–48. DOI: 10.1016/j.strueco.2023.02.004

Carlo, R. (2024, June 27). Marialuisa, Menanno., Ilenia, Zennaro., Matteo, Mario, Savino. (2024). 3. A New Methodological Framework for Optimizing Predictive Maintenance Using Machine Learning Combined with Product Quality Parameters. *Machines (Basel), 12*(7), 443. Advance online publication. DOI: 10.3390/machines12070443

Chatterjee, A., & Gupta, S. (2023). *Machine Learning for Quality Control in Aerospace Engineering*. CRC Press.

Chebbi, J., Defaÿ, F., Brière, Y., & Deruaz-Pepin, A. (2020, April). Novel model-based control mixing strategy for a coaxial push-pull multirotor. *IEEE Robotics and Automation Letters, 5*(2), 485–491. DOI: 10.1109/LRA.2019.2963652

Cheng, G., Han, J., & Lu, X. (2017). Remote Sensing Image Scene Classification: Benchmark and State of the Art. *Proceedings of the IEEE, 105*(10), 1865–1883. DOI: 10.1109/JPROC.2017.2675998

Chen, L., & Zhao, Q. (2021). Deep learning applications in aerospace manufacturing quality control. *Journal of Intelligent Systems, 28*(4), 456–467.

ComplianceQuest. (2022). AI in quality management solutions for the aerospace industry

Components, B. (2025). AI in Quality Control and Compliance in Aerospace. *Broadline Components White Paper.*

Components, B. (n.d.). AI for quality control and compliance in aerospace sourcing. Broadline Components., Retrieved March 28, 2025, from.

Concept, N. (2025). Aerospace parts manufacturing and AI: The efficiency guide. Neural Concept., Retrieved March 28, 2025, from.

Cooley, C. H. (1922). *Human nature and the social order.* Scribner.

Core Business Solutions. (2025). *AS9100 Aerospace Quality Management System. International Aerospace Quality Group.* IAQG.

Dadhwal, V. K., & Parihar, J. S. (2020). Space-based crop monitoring in India: The FASAL experience. *Current Science, 119*(6), 940–949.

Dangeti, C. S. (2018). Sensitivity of Turbulence Closure Coefficients on the Aerodynamic Predictions of Flow over a Simplified Road Vehicle (Master's thesis, The University of North Carolina at Charlotte).

Dasari, A., Nguyen, P., & Kumar, S. (2020). Quality assurance in additive manufacturing: Defect classification using random forest algorithms. *Additive Manufacturing, 34*, 101234.

Dasari, K., Mekala, S., & Kaka, J. R. (2024). Evaluation of UDP-Based DDoS Attack Detection by Neural Network Classifier with Convex Optimization and Activation Functions. *Ingénierie des Systèmes d'Information, 29*(3), 3. DOI: 10.18280/isi.290321

Deep Learning for Visual Inspection in Aerospace. (2024)... *IEEE Transactions on Industrial Informatics, 60*(1), 25–40.

Deliencourt, B., Giraud, M., Miranda, J., Larnier, S., & Herbulot, A. (n.d.). *Génération photo-réaliste de défauts sur des images de surfaces extérieures d'avions acquises par un drone autonome*. ORASIS, Journées Francophones des Jeunes Chercheurs en Vision Par Ordinateur.

Deng, L., Liu, G., & Zhang, Y. (2024, April). A Review of Machine Vision Applications in Aerospace Manufacturing Quality Inspection. In 2024 4th International Conference on Computer, Control and Robotics (ICCCR) (pp. 31-39). IEEE. DOI: 10.1109/ICCCR61138.2024.10585378

Dodin, P., Xiao, J., Adulyasak, Y., Alamdari, N. E., Gauthier, L., Grangier, P., Lemaitre, P., & Hamilton, W. L. (2023). Bombardier Aftermarket Demand Forecast with Machine Learning. *INFORMS Journal on Applied Analytics*, *53*(6), 425–445. DOI: 10.1287/inte.2023.1164

Donecle. (n.d.). In *Wikipedia*. Retrieved March 28, 2025

Du, X., Martins, J. R., O'Leary-Roseberry, T., Chaudhuri, A., Ghattas, O., & Willcox, K. E. (2023). Learning optimal aerodynamic designs through multi-fidelity reduced-dimensional neural networks. In AIAA SCITECH 2023 Forum (p. 0334). DOI: 10.2514/6.2023-0334

Dube, P., & Hiravennavar, S. (2020). Machine Learning Approach to Predict Aerodynamic Performance of Underhood and Underbody Drag Enablers. In SAE Technical Papers (Vol. 2020-April). SAE International. https://doi.org/DOI: 10.4271/2020-01-0684

Dube, A. (2024). Application of Deep Learning in Predictive Maintenance of Aircraft Engines. *Darpan International Research Analysis*, *12*(3), 83–100. DOI: 10.36676/dira.v12.i3.58

Dush, D. M., Hirt, M. L., & Schroeder, H. E. (1983). Self-statement modification with adults: A meta-analysis. *Psychological Bulletin*, *94*(3), 408–422. DOI: 10.1037/0033-2909.94.3.408

D-Wave. (2022). *Quantum computing for aerospace*. https://www.dwavesys.com

Dwivedi, K., Govindarajan, P., Srinivasan, D., Keerthi Sanjana, A., Selvanambi, R., & Karuppiah, M. (2023). Intelligent Autonomous Drones in Industry 4.0. In Sarveshwaran, V., Chen, J. I. Z., & Pelusi, D. (Eds.), *Artificial Intelligence and Cyber Security in Industry 4.0. Advanced Technologies and Societal Change*. Springer., DOI: 10.1007/978-981-99-2115-7_6

Eagle Technologies. (2022). 5 applications for advanced vision systems in aerospace.

EASA Guidelines on AI-Based Quality Inspection. (2023). *European Union Aviation Safety Agency*. EASA.

Eivazi, H., Le Clainche, S., Hoyas, S., & Vinuesa, R. (2022). Towards extraction of orthogonal and parsimonious non-linear modes from turbulent flows. *Expert Systems with Applications*, *202*, 117038. DOI: 10.1016/j.eswa.2022.117038

Elham, A. (2015). Adjoint quasi-three-dimensional aerodynamic solver for multi-fidelity wing aerodynamic shape optimization. *Aerospace Science and Technology*, *41*, 241–249. DOI: 10.1016/j.ast.2014.12.024

English, S., Sabatine, J. M., & Brownell, P. (2019). *Professional coaching: Principles and practice*. Springer Publishing Company, LLC.

Erdös, T., de Haan, E., & Heusinkveld, S. (2020). Coaching: Client factors & contextual dynamics in the change process. *Coaching (Abingdon, UK)*, *14*(2), 162–183. DOI: 10.1080/17521882.2020.1791195

Ertugrul, G., Alimov, A., Sviridov, A., & Härtel, S. Machine learning application for optimization of laser directed energy deposition process for aerospace component rapid prototyping in additive manufacturing. Materials Research Proceedings, 41.

Eskandariyun, A., Fu, H., Stere, A., Bauer, A., Reynolds, C., Chengalva, M., . . . Zobeiry, N. (2024, April). A Novel Machine Learning Framework for Digital Estimation of Allowables in Aerospace Composites. In ASME Aerospace Structures, Structural Dynamics, and Materials Conference (Vol. 87745, p. V001T01A014). American Society of Mechanical Engineers. DOI: 10.1115/SSDM2024-121597

Eslami, S. M. A., Heess, N., Weber, T., Tassa, Y., Szepesvari, D., Kavukcuoglu, K., & Hinton, G. E. “Attend, Infer, Repeat: Fast Scene Understanding with Generative Models”, *30th Conference on Neural Information Processing Systems (NIPS 2016)*, Barcelona, Spain, 2016.

European Commission. (2020). *Copernicus Open Access Hub*. https://scihub.copernicus.eu

European Commission. (2020). *General Data Protection Regulation (GDPR)*. https://ec.europa.eu

European Union Aviation Safety Agency (EASA). (2022). *AI roadmap: A human-centric approach to AI in aviation*. https://www.easa.europa.eu

FAA AI Integration for Aerospace Safety. (2024). *Federal Aviation Administration*. FAA.

Falcão e Cunha, N., Gan, T. S., Curcio, E., Amorim, P., Almada-Lobo, B., & Grunow, M. (2023). Robust supply chain design with suppliers as system integrators: An aerospace case study. *International Journal of Production Research, 61*(15), 5244–5265. DOI: 10.1080/00207543.2022.2099769

Farnell, C., Jackson, J., Corbitt, A., & Mantooth, H. A. (2023, September). Rapid prototyping of a SiC-based PMSM motor drive for aerospace applications. In *2023 IEEE Design Methodologies Conference (DMC)* (pp. 1-5). IEEE. DOI: 10.1109/DMC58182.2023.10412479

Ferrero, M., Vadillo, M. A., & León, S. P. (2021). A valid evaluation of the theory of multiple intelligences is not yet possible: Problems of methodological quality for intervention studies. *Intelligence, 88*, 101566. DOI: 10.1016/j.intell.2021.101566

Flyvbjerg, B. (2011). Case Study. In N. K. Denzin & Y. S. Lincoln (Eds.), *The SAGE Handbook of Qualitative Research* (pp. 301–316). essay.

Frey Marioni, Y., de Toledo Ortiz, E. A., Cassinelli, A., Montomoli, F., Adami, P., & Vazquez, R. (2021). A machine learning approach to improve turbulence modelling from DNS data using neural networks. International Journal of Turbomachinery. *Propulsion and Power, 6*(2), 17.

Fu, C., Uddin, M., Robinson, C., Guzman, A., & Bailey, D. (2017). Turbulence models and model closure coefficients sensitivity of NASCAR Racecar RANS CFD aerodynamic predictions. SAE International Journal of Passenger Cars-Mechanical Systems, 10(2017-01-1547), 330-344.

Gao, Y. (2023). CFD based aerodynamic optimization design. *Theoretical and Natural Science, 18*(1), 83–90. DOI: 10.54254/2753-8818/18/20230343

Gardner, H., & Hatch, T. (1989). Educational implications of the theory of multiple intelligences. [CrossRef] [Google Scholar]. *Educational Researcher, 18*(8), 4–10. DOI: 10.3102/0013189X018008004

Genevaux, G., Daniel, L., & Chevalier, T. (2022). A digital twin approach for predictive maintenance of aircraft systems.

Gill, N. S. (2024). AI-driven quality control: Revolutionizing manufacturing standards. Xenon Stack., Retrieved March 28, 2025, from.

Glaessgen, E., & Stargel, D. (2012). The digital twin paradigm for future NASA and US Air Force vehicles.

Godunov, S. K., & Bohachevsky, I. (1959). Finite difference method for numerical computation of discontinuous solutions of the equations of fluid dynamics. Matematičeskij sbornik, 47(3), 271-306.

Goksidan, H. T. (2019). Value Creation in Project Management: The Effect of Project Lifecycle Management Measures in an Aerospace Company. Strategica: Upscaling Digital. *Transformations in Business & Economics*, (July), 724–734.

Goleman, D. (1998). *Working with Emotional Intelligence*. Bantam.

Goleman, D. (2007). *Emotional Intelligence* (10th ed.). Bantam Books.

Gómez-Carmona, O., Casado-Mansilla, D., López-de-Ipiña, D., & García-Zubia, J. (2024). Human-in-the-loop machine learning: Reconceptualizing the role of the user in interactive approaches. *Internet of Things : Engineering Cyber Physical Human Systems, 25*, 101048. DOI: 10.1016/j.iot.2023.101048

Goodfellow, I., Bengio, Y., & Courville, A. (2016). *Deep learning*. MIT Press.

Govindarajan, B., Jaganraj, R., Srinivasan, V., & Kumaran, T. (2025). Evolution of UAV Technology From Early Innovations to Future Horizons. In *Innovations and Developments in Unmanned Aerial Vehicles* (pp. 53–94). IGI Global Scientific Publishing.

Graham-Knight, B., Körber, M., Djavadifar, A., & Najjaran, H. (2021). Automated visual detection of geometrical defects in composite manufacturing processes using deep convolutional neural networks. *Journal of Intelligent Manufacturing*, *32*(5), 1235–1248.

Grant, A. M. (2003). The impact of life coaching on Goal Attainment, metacognition and Mental Health. *Social Behavior and Personality*, *31*(3), 253–263. DOI: 10.2224/sbp.2003.31.3.253

Guilmineau, E., Deng, G. B., Leroyer, A., Queutey, P., Visonneau, M., & Wackers, J. (2018). Assessment of hybrid RANS-LES formulations for flow simulation around the Ahmed body. *Computers & Fluids*, *176*, 302–319. DOI: 10.1016/j.compfluid.2017.01.005

Habib, M. K., & Mohamed, K. (2024, August). Enhancing Predictive Maintenance Hyperparameter Optimization and Adopted Strategies. In *2024 IEEE International Conference on Mechatronics and Automation (ICMA)* (pp. 153-158). IEEE. DOI: 10.1109/ICMA61710.2024.10633049

Hajela, P., Sakalkar, V., & Mullur, A. (2009). Multidisciplinary Analysis and Design Tools For Uncertainty Modeling. *Journal of Aerospace Sciences and Technologies*, •••, 240–251.

Hanover, D., Loquercio, A., Bauersfeld, L., Romero, A., Penicka, R., Song, Y., Cioffi, G., Kaufmann, E., & Scaramuzza, D. (2024). Autonomous drone racing: A survey. *IEEE Transactions on Robotics*, *20*, 3044–3067. DOI: 10.1109/TRO.2024.3400838

Hardy, J. (2006). Speaking clearly: A critical review of the self-talk literature. *Psychology of Sport and Exercise*, *7*(1), 81–97. DOI: 10.1016/j.psychsport.2005.04.002

Hariharan, B., & Ross, B. Girshick, “Low-Shot Visual Recognition by Shrinking and Hallucinating Features”, *International Conference on Computer Vision (ICCV)*, pp. 3037-3046, 2017.

Hartl, D. J., Mabe, J. H., Benafan, O., Coda, A., Conduit, B., Padan, R., & Doren, B. V. (2015). Standardization of shape memory alloy test methods toward certification of aerospace applications. *Smart Materials and Structures, 24*(8), 082001. Advance online publication. DOI: 10.1088/0964-1726/24/8/082001

Hastie, T., Tibshirani, R., & Friedman, J. (2009). *The elements of statistical learning: Data mining, inference, and prediction*. Springer. DOI: 10.1007/978-0-387-84858-7

Higgins, E. T. (1989). Self-discrepancy theory: What patterns of self-beliefs cause people to suffer? *Advances in Experimental Social Psychology, 22*, 93–136. DOI: 10.1016/S0065-2601(08)60306-8

Honeycutt, J. M., & Ford, S. G. (2001). Mental imagery and Intrapersonal Communication: A review of research on imagined interactions (IIS) and current developments. *Annals of the International Communication Association, 25*(1), 315–345. DOI: 10.1080/23808985.2001.11679007

Hua, Y., Mou, L., & Zhu, X. X. (2019). Relation Network for Multi-label Aerial Image Classification. *arXiv preprint arXiv:1907.07274*.

Humfeld, K. D., Gu, D., Butler, G. A., Nelson, K., & Zobeiry, N. (2021). A machine learning framework for real-time inverse modeling and multi-objective process optimization of composites for active manufacturing control. *arXiv preprint arXiv:2104.11342*.

Humml, J. M., Oshima, E., O’Gara, S., Rusch, A., Gharib, M., Lee, V., & Khodadoust, A. (2024). Development of machine learning tools for aerospace design: wind tunnel investigations on a speed bump model. In AIAA SCITECH 2024 Forum (p. 2834). Ilie, M., & Semenescu, A. Supply Chain Management Of Manufacturing Processes Using Machine Learning Technique. DOI: 10.2514/6.2024-2834

ICAO. (2021). *Cybersecurity in civil aviation: A global perspective*. International Civil Aviation Organization.

InovarTech. (2022). Enhancing worker safety & compliance with AI vision in manufacturing.

International Civil Aviation Organization (ICAO). (2021). *Manual on machine learning for aviation safety*. ICAO Publishing.

ISO 9001: Quality Management Systems for Aerospace. (2022). *International Organization for Standardization (ISO).*

Jeon, J., Lee, J., & Kim, S. J. (2022). Finite volume method network for the acceleration of unsteady computational fluid dynamics: Non-reacting and reacting flows. *International Journal of Energy Research, 46*(8), 10770–10795. DOI: 10.1002/er.7879

Jiang, Y., Tran, T. H., & Williams, L. (2023). Machine learning and mixed reality for smart aviation: Applications and challenges. *Journal of Air Transport Management, 111*, 102437. Advance online publication. DOI: 10.1016/j.jairtraman.2023.102437

Jordan, M., & Livingstone, J. B. (2013, July). *Coaching vs psychotherapy in health and wellness: Overlap, dissimilarities, and the potential for collaboration*. Global advances in health and medicine. https://www.ncbi.nlm.nih.gov/pmc/articles/PMC3833547/

Jordan, M. I., & Mitchell, T. M. (2015). Machine learning: Trends, perspectives, and prospects. *Science, 349*(6245), 255–260. DOI: 10.1126/science.aaa8415 PMID: 26185243

Jose, L. A.Jr, Brintrup, A., & Salonitis, K. (2020). Analysing the evolution of aerospace ecosystem development. *PLoS One, 15*(4), e0231985. DOI: 10.1371/journal.pone.0231985 PMID: 32343729

Kalyan, B. S. (2024, February). A multi-fidelity platform using machine learning for integrated design of aircraft systems. In *2024 International Conference on Emerging Systems and Intelligent Computing (ESIC)* (pp. 243-248). IEEE. DOI: 10.1109/ESIC60604.2024.10481666

Karimzadeh, M., Basvoju, D., Vakanski, A., Charit, I., Xu, F., & Zhang, X. (2024). Machine Learning for Additive Manufacturing of Functionally Graded Materials. *Materials (Basel), 17*(15), 3673. DOI: 10.3390/ma17153673 PMID: 39124337

Karniadakis, G. E., Kevrekidis, I. G., Lu, L., Perdikaris, P., Wang, S., & Yang, L. (2021). Physics-informed machine learning. *Nature Reviews. Physics*, *3*(6), 422–440. DOI: 10.1038/s42254-021-00314-5

Katkuri, A. V. R., Madan, H., Khatri, N., Abdul-Qawy, A. S. H., & Patnaik, K. S. (2024). Autonomous UAV Navigation using Deep Learning-Based Computer Vision Frameworks: A Systematic Literature Review. *Array (New York, N.Y.), 100361*, 100361. Advance online publication. DOI: 10.1016/j.array.2024.100361

Kato, Y. (2025). Balancing Cost and Quality in. Aerospace Manufacturing with AI. *Quality*, *64*(4), 32–32.

Kochenderfer, M. J., & Wheeler, T. A. (2019). *Algorithms for decision making*. MIT Press.

Konovalenko, I., Kuznetsova, E., Miller, A., Miller, B., Popov, A., Shepelev, D., & Stepanyan, K. (2018). New approaches to the integration of navigation systems for autonomous unmanned vehicles (UAV). *Sensors (Basel), 18*(9), 3010. DOI: 10.3390/s18093010 PMID: 30205568

Kraichnan, R. H. (1967). Inertial ranges in two-dimensional turbulence. *The Physics of Fluids*, *10*(7), 1417–1423. DOI: 10.1063/1.1762301

Kuhn, T. (2002). Negotiating boundaries between scholars and Practitioners. *Management Communication Quarterly*, *16*(1), 106–112. DOI: 10.1177/0893318902161008

Kumar, R., & Sharma, P. (2020). Predictive maintenance in aerospace engineering through machine learning. *Journal of Aerospace Operations*, *8*(2), 123–135.

Kumar, S., Patel, D., & Singh, R. P. (2023). *Thermodynamic Radiators: Principles*. Applications, and Performance Analysis.

Laker, B. (2023, July 21). *Every leader can benefit from coaching. Here's why...* Forbes. https://www.forbes.com/sites/benjaminlaker/2022/10/04/every-leader-can-benefit-from-coaching-heres-why/

LeCun, Y., Bengio, Y., & Hinton, G. (2015). Deep learning. *Nature, 521*(7553), 436–444. DOI: 10.1038/nature14539 PMID: 26017442

Lee, K., & Wong, M. (2022). Computer vision techniques for real-time aircraft inspection. *International Journal of Aviation Engineering, 39*(2), 123–134.

Lee, S. C. (2023). A Literary Survey on the Challenges of Adopting Lean in Aerospace Production and Aviation Maintenance. *Journal of Aviation Technology and Engineering, 12*(1). Advance online publication. DOI: 10.7771/2159-6670.1243

Leonard, C., Prasetiyo, E. E., & Putra, I. R. (2023). Design and implementation of electrical system morphing wing flight control on prototype light aircraft. *International Journal of Power Electronics and Drive Systems, 14*(2), 781–788. DOI: 10.11591/ijpeds.v14.i2.pp781-788

Ling, J., Kurzawski, A., & Templeton, J. (2016). Reynolds averaged turbulence modelling using deep neural networks with embedded invariance. *Journal of Fluid Mechanics, 807*, 155–166. DOI: 10.1017/jfm.2016.615

Li, P., Chen, P., & Zhang, D. (2022). Cross-Modal Feature Representation Learning and Label Graph Mining in a Residual Multi-Attentional CNN-LSTM Network for Multi-Label Aerial Scene Classification. *Remote Sensing (Basel), 14*(10), 2424. DOI: 10.3390/rs14102424

Liu, B., Cui, Z., Hu, S., Li, X., Lin, H., & Zhang, Z. (2025). LLM evaluation based on aerospace manufacturing expertise: Automated generation and multi-model question answering. *arXiv preprint arXiv:2501.17183*

Liu, W., Fang, J., Rolfo, S., Moulinec, C., & Emerson, D. R. (2021). An iterative machine-learning framework for RANS turbulence modelling. *International Journal of Heat and Fluid Flow, 90*, 108822. DOI: 10.1016/j.ijheatfluidflow.2021.108822

Liu, Z., Mao, H., Wu, C.-Y., Feichtenhofer, C., Darrell, T., & Xie, S. (2022). A ConvNet for the 2020s. *Proceedings of the IEEE/CVF Conference on Computer Vision and Pattern Recognition (CVPR)*, 11976–11986. https://doi.org/DOI: 10.1109/CVPR52688.2022.01170

Li, Y., Chen, R., Zhang, Y., Zhang, M., & Chen, L. (2020). Multi-label remote sensing image scene classification by combining a convolutional neural network and a graph neural network. *Remote Sensing (Basel)*, *12*(23), 4003. DOI: 10.3390/rs12234003

Luft, J. A., Jeong, S., Idsardi, R., & Gardner, G. (2022). Literature Reviews, Theoretical Frameworks, and Conceptual Frameworks: An Introduction for New Biology Education Researchers. *CBE Life Sciences Education*, *21*(3), rm33. DOI: 10.1187/cbe.21-05-0134 PMID: 35759629

Maghami, A., Salehi, M., & Khoshdarregi, M. (2021). Automated vision-based inspection of drilled CFRP composites using multi-light imaging and deep learning. *Journal of Manufacturing Processes*, *64*, 123–134.

Martinez, R., & Gonzalez, L. (2023). Enhancing defect detection in aerospace composites using convolutional neural networks. *Composite Structures*, *250*, 112345.

Maulik, R., San, O., Rasheed, A., & Vedula, P. (2019). Subgrid modelling for two-dimensional turbulence using neural networks. *Journal of Fluid Mechanics*, *858*, 122–144. DOI: 10.1017/jfm.2018.770

Mayer, J. D., Salovey, P., & Caruso, D. (2004). Emotional intelligence: Theory, findings and implications. *Psychological Inquiry*, *15*(3), 197–215. DOI: 10.1207/s15327965pli1503_02

McGarry, L., Butterfield, J., Murphy, A., & Higgins, C. (2023). Machine learning methods to improve the accuracy of industrial robots. SAE International Journal of Advances and Current Practices in Mobility, 5(2023-01-1000), 1900-1918.

Meister, R., Chen, Y., & Wang, Z. (2021). Integrating CNN and SVM models for enhanced visual inspection in aerospace applications. *IEEE Transactions on Industrial Informatics*, *17*(6), 4321–4330.

Merola, S., Guida, M., & Marulo, F. Digital optics and machine learning algorithms for aircraft maintenance. Materials Research Proceedings, 42.

Mingyu, L. (2024, August). Haotian, Li., Hongyuan, Zhou., Hong, Zhang., Guangyan, Huang. (2024). 1. Development of machine learning methods for mechanical problems associated with fibre composite materials: A review. *Composites Communications*, *49*, 101988. Advance online publication. DOI: 10.1016/j.coco.2024.101988

Ministry of Education. (2021). *Annual report on technical education and skill development in India*. Government of India.

Miranda, J., Larnier, S., & Claybrought, M. (n.d.). Caractérisation d'objets sur des images acquises par drone recherchant des défauts pour la maintenance aéronautique. *Actes de la conférence Reconnaissance de Formes et Intelligence Artificielle.*

Miranda, J., Larnier, S., & Claybrought, M. (2020). AI-based defect detection for aircraft maintenance. *Journal of Aerospace Engineering*, *37*(4), 56–72.

Mishra, R., Ranjan Kumar, H., Nayak, S., Nayak, S. K., Mishra, S. B., & Nanda, B. K. (2022, November). Structural Analysis of Aircraft Wing Considering Titanium Alloy. In International conference on Advances in Materials and Manufacturing (pp. 171-178). Singapore: Springer Nature Singapore.

Mohammadi, A., Al-Fuqaha, A., Sorour, S., & Guizani, M. (2018). Deep learning for IoT big data and streaming analytics: A survey. *IEEE Communications Surveys and Tutorials*, *20*(4), 2923–2960. DOI: 10.1109/COMST.2018.2844341

Morand, L., Iraki, T., Dornheim, J., Sandfeld, S., Link, N., & Helm, D. (2023). Machine learning for structure-guided materials and process design. arXiv preprint arXiv:2312.14552.

Morita, Y., Rezaeiravesh, S., Tabatabaei, N., Vinuesa, R., Fukagata, K., & Schlatter, P. (2022). Applying Bayesian optimization with Gaussian process regression to computational fluid dynamics problems. *Journal of Computational Physics*, *449*, 110788. DOI: 10.1016/j.jcp.2021.110788

Mudraje, A., Moser, B. B., Frolov, S., & Dengel, A. "Multi-label scene classification in remote sensing benefits from image super-resolution," *arXiv preprint arXiv:2501.06720*, Jan. 2025.

Muelaner, J. E. (2024). Generative Design in Aerospace and Automotive Structures (No. EPR2024016). SAE Technical Paper.

Mumtaz, A. Shah. (2015). 6. Commercial Aerospace Industry from the U.S. Monopoly to Duopoly. Social Science Research Network, doi: DOI: 10.2139/SSRN.2545501

Nahavandi, S., Alizadehsani, R., Nahavandi, D., Lim, C. P., Kelly, K., & Bello, F. (2024). Machine learning meets advanced robotic manipulation. *Information Fusion*, *105*, 102221. DOI: 10.1016/j.inffus.2023.102221

NASA. (2021). *Ingenuity Mars Helicopter – Technology Demonstration*. https://mars.nasa.gov

Nascimento, D. L., Silva, R. G., & Martins, M. F. (2021). Machine learning applications in aircraft structural health monitoring. *Journal of Aerospace Technology and Management*, *13*(1), e2021–e1213. DOI: 10.5028/jatm.c2021-1213

Nguyen, T., & Tran, P. (2020). Automated visual inspection systems in aircraft maintenance. *Aerospace Maintenance Journal*, *15*(1), 89–99.

Ng, W. L., Goh, G. L., Goh, G. D., Ten, J. S. J., & Yeong, W. Y. (2024). Progress and opportunities for machine learning in materials and processes of additive manufacturing. *Advanced Materials*, *36*(34), 2310006. DOI: 10.1002/adma.202310006 PMID: 38456831

Ni, Z. (2022). Reframe the Field of Aerospace Engineering Via Machine Learning: Application and Comparison. In Journal of Physics: Conference Series (Vol. 2386). Institute of Physics. https://doi.org/DOI: 10.1088/1742-6596/2386/1/012031

Nicolau, A., Candel, O. S., Constantin, T., & Kleingeld, A. (2023). The effects of executive coaching on behaviors, attitudes, and personal characteristics: A meta-analysis of Randomized Control Trial Studies. *Frontiers in Psychology*, *14*, 1089797. Advance online publication. DOI: 10.3389/fpsyg.2023.1089797 PMID: 37333584

Nogueira, R. P. R., Melicio, R., Valério, D., & Santos, L. F. F. M. (2023). Learning Methods and Predictive Modeling to Identify Failure by Human Factors in the Aviation Industry. *Applied Sciences (Basel, Switzerland)*, *13*(6), 4069. Advance online publication. DOI: 10.3390/app13064069

Obiols-Sales, O., Vishnu, A., Malaya, N., & Chandramowliswharan, A. (2020, June). CFDNet: A deep learning-based accelerator for fluid simulations. In *Proceedings of the 34th ACM international conference on supercomputing* (pp. 1-12). DOI: 10.1145/3392717.3392772

OECD. (2020). *The future of the space economy: Trends, challenges and growth opportunities*. OECD Publishing., DOI: 10.1787/c5996201-

Office for Human Research Protections (OHRP). (2023, May 31). *Exemptions (2018 requirements)*. Retrieved from https://www.hhs.gov/ohrp/regulations-and-policy/regulations/45-cfr-46/common-rule-subpart-a-46104/index.html

Okafor, E. O., & Nwosu, C. N. (2020). Machine learning in autonomous UAV navigation: Trends and future directions. *Journal of Intelligent & Robotic Systems*, *99*(3), 421–435. DOI: 10.1007/s10846-020-01215-8

Olsen, J. M., & Nesbitt, B. J. (2010). Health coaching to improve healthy lifestyle behaviors: An integrative review. *American Journal of Health Promotion*, *25*(1), e1–e12. Advance online publication. DOI: 10.4278/ajhp.090313-LIT-101 PMID: 20809820

Omdena (2023). Edge AI and Real-Time Defect Detection. *International Journal of Manufacturing Science, 12*(3), 210-230.

Padmavathi, A., & Vyasasreevedansh, B. (2024) Smart Surveillence Drone with Navigation System, *2024 IEEE International Conference on Computing, Power and Communication Technologies (IC2PCT)*, Greater Noida, India, *5*, 1452-1458, https://doi.org/DOI: 10.1109/IC2PCT60090.2024.10486563

Paleja, R., Munje, M., Chang, K., Jensen, R., & Gombolay, M. (2024). Designs for Enabling Collaboration in Human-Machine Teaming via Interactive and Explainable Systems. arXiv preprint arXiv:2406.05003.

Passmore, J. (2020). Coaching defined and explored. In *The coaches' handbook: The complete practitioner guide for professional coaches* (First). essay, Routledge. DOI: 10.4324/9781003089889-2

Patel, D. (2023). Exploring the Frontiers of Microfluidics: Challenges and Future Prospects. *Advances in MEMS and Microfluidic Systems*, 11-31.

Patel, D. (2024). Sustainable Renewable Energy Sources and Emerging Technologies. *Optimization Techniques for Hybrid Power Systems: Renewable Energy, Electric Vehicles, and Smart Grid*, 343-361.

Patel, D. (2025). Emerging Sustainable Materials to Improve Green Energy: Environmental Applications and Future Scope. In *Innovations in Energy Efficient Construction Through Sustainable Materials* (pp. 65-82). IGI Global.

Patel, D. K. (2018). Evaluation of Perforance of IC Engine using Alternate Fuel.

Patel, D., & Mishra, A. (2023). Hybrid sustainable nanomaterials using for nanofluids of advance applications and challenges of future scope.

Patel, D., Mishra, A., & Nabeel, M. (2022, February). Heat transfer characteristics of nanofluids of silicon oxides (SiO2) with conventional fluid. In *2022 2nd International Conference on Innovative Practices in Technology and Management (ICIPTM)* (Vol. 2, pp. 420-423). IEEE.

Patel, D., Singh, R. P., Rajput, R. S., & Tiwari, P. (2022, December). Thermophyscical Properties And Applications Nanofluids–On Review. In *2022 5th International Conference on Contemporary Computing and Informatics (IC3I)* (pp. 1324-1328). IEEE. DOI: 10.1109/IC3I56241.2022.10072842

Patel, D. (2025). Emerging sustainable nanomaterials and their applications and future scope. In *Advances in Sustainable Materials* (pp. 107–135). Elsevier. DOI: 10.1016/B978-0-443-13849-2.00005-3

Patel, D., Sharma, A., & Mishra, A. (2021, November). Study of convective heat transfer characteristics of nano fluids in circular tube. In *2021 International Conference on Technological Advancements and Innovations (ICTAI)* (pp. 264-267). IEEE. DOI: 10.1109/ICTAI53825.2021.9673432

Patel, D., Singh, A. K., Sarda, C., & Raj, R. (2017). Review Six Stroke Engine. [IJCI]. *International Journal on Cybernetics & Informatics*, *6*(2), 53.

Patel, D., Varma, I. P., & Khan, F. A. (2022). A Review Advanced Vehicle with Automatic Pneumatic Bumper System using Two Cylinder. *Int. Res. J. Eng. Tech*, *9*, 264–267.

Patil, A. (2023). The good, the bad, and the awful of AI in aerospace. Aerospace Manufacturing and Design., Retrieved March 28, 2025, from.

Piaget, J. (1923). *Language and Thought of the Child.* Routledge.

Pope, S. B. (1999). A perspective on turbulence modelling. In *Modeling complex turbulent flows* (pp. 53–67). Springer Netherlands. DOI: 10.1007/978-94-011-4724-8_5

Pop, G. I., Titu, A. M., & Pop, A. B. (2023). Enhancing Aerospace Industry Efficiency and Sustainability: Process Integration and Quality Management in the Context of Industry 4.0. *Sustainability (Basel)*, *15*(23), 16206. DOI: 10.3390/su152316206

Poroseva, S. V., Colmenares, F. J. D., & Murman, S. M. (2016). On the accuracy of RANS simulations with DNS data. *Physics of Fluids*, *28*(11), 115102. DOI: 10.1063/1.4966639 PMID: 30271109

Posner, P. L. (2009). The pracademic: An agenda for re-engaging practitioners and academics. *Public Budgeting &amp. Public Budgeting & Finance*, *29*(1), 12–26. DOI: 10.1111/j.1540-5850.2009.00921.x

Poudel, L., Jha, S., Meeker, R., ... (2024). Predictive quality assessment for ultrasonic additive manufacturing with deep learning. *arXiv preprint arXiv:2410.24055.*

Poudel, L., Jha, S., Meeker, R., Phan, D.-N., & Bhowmik, R. (2024). Advanced predictive quality assessment for ultrasonic additive manufacturing with deep learning model. *arXiv preprint arXiv:2410.24055.*

Preskill, J. (2018). Quantum computing in the NISQ era and beyond. *Quantum : the Open Journal for Quantum Science*, *2*, 79. DOI: 10.22331/q-2018-08-06-79

Qian, B., Al Said, N., & Dong, B. (2024). New technologies for UAV navigation with real-time pattern recognition. *Ain Shams Engineering Journal*, *15*(3), 102480. DOI: 10.1016/j.asej.2023.102480

Qi, X., Zhu, P., Wang, Y., Zhang, L., Peng, J., Wu, M., Chen, J., Zhao, X., Zang, N., & Mathiopoulos, P. T. (2020). MLRSNet: A multi-label high spatial resolution remote sensing dataset for semantic scene understanding. *ISPRS Journal of Photogrammetry and Remote Sensing*, *169*, 337–350. DOI: 10.1016/j.isprsjprs.2020.09.020

Quest, C. (n.d.). AI in quality management solutions for the aerospace industry. Compliance Quest., Retrieved March 28, 2025, from.

Rabault, J., Kuchta, M., Jensen, A., Réglade, U., & Cerardi, N. (2019). Artificial neural networks trained through deep reinforcement learning for active flow control.

Raes, F., Pommier, E., Neff, K. D., & Van Gucht, D. (2011). Construction and factorial validation of a short form of the Self-Compassion Scale. *Clinical Psychology & Psychotherapy*, *18*(3), 250–255. DOI: 10.1002/cpp.702 PMID: 21584907

Raja, S., Al-Tmimi, H. M., Ghadir, G. K., Mustafa, M. A., Alani, Z. K., Rusho, M. A., & Rajeswari, N. (2024). An analysis of polymer material selection and design optimization to improve Structural Integrity in 3D printed aerospace components. Applied Chemical Engineering, 1875-1875.

Rajput, R. S., Patel, D., & Singh, R. P. (2023). Applications and Feasibility of Large-Scale Solar-Powered Peltier Refrigeration Systems.

Rangel, P., Tardif, S., Mehrubeoglu, M., John, E. S., Whaley, P., Salas, M., . . . Torres, M. (2024, June). IoT-enabled unmanned traffic management system with dynamic vision-based drone detection for sense and avoid coordination. In *Real-Time Image Processing and Deep Learning 2024* Orlando, Florida, US, SPIE *13034*, 99-111 https://doi.org/DOI: 10.1117/12.3014009

Ravindran, S. (2022). The rise of private space in India. *Nature India, 18*, 324–325. DOI: 10.1038/nindia.2022.140

ResearchGate. (2022). A review of machine vision applications in aerospace manufacturing quality inspection.

Roberts, C. V. (1983, October 31). *The definition and delimitation of intrapersonal communication: A physiological perspective.* ERIC. https://eric.ed.gov/?id=ED240634

Roberts, J. (2024). *Computer Vision in Manufacturing: Advanced Defect Detection Techniques*. Wiley.

Rodriguez, A. A., Shekaramiz, M., & Masoum, M. A. (2024). Computer Vision-Based Path Planning with Indoor Low-Cost Autonomous Drones: An Educational Surrogate Project for Autonomous Wind Farm Navigation. *Drones (Basel)*, *8*(4), 154. DOI: 10.3390/drones8040154

Root, Al (2024). GE partners with Microsoft to bring new AI tools to its workforce. *Barron's*. Retrieved March

Root, Al (2025). GE partners with Microsoft to bring new AI tools to its workforce. *Barron's*. Retrieved March 28, 2025

Rosell, A., Svenman, E., Westphal, P., Mukundan, A., Bhattacharya, S., Bharthulwar, S., . . . Jhanardhanan, S. (2023, September). Machine learning-based system to automate visual inspection in aerospace engine manufacturing. In 2023 IEEE 28th International Conference on Emerging Technologies and Factory Automation (ETFA) (pp. 1-8). IEEE. DOI: 10.1109/ETFA54631.2023.10275515

Sadrehaghighi, Ideen. (2022). Adaptive Meshing.

SAE ARP 6297: AI-Based Quality Control in Aerospace Manufacturing. (2024). *Society of Automotive Engineers (SAE).*

Sarojini, D., & Mavris, D. (2022). Structural Analysis and Optimization of Wings Subjected to Dynamic Loads. *AIAA Journal*, *60*(2), 1013–1023. DOI: 10.2514/1.J060931

Schmidhuber, J. (2015). Deep learning in neural networks: An overview. *Neural Networks*, *61*, 85–117. DOI: 10.1016/j.neunet.2014.09.003 PMID: 25462637

Schoenholz, C., & Zobeiry, N. (2024). An accelerated process optimization method to minimize deformations in composites using theory-guided probabilistic machine learning. *Composites. Part A, Applied Science and Manufacturing*, *176*, 107842. DOI: 10.1016/j.compositesa.2023.107842

Schutte, N. S., Malouff, J. M., Hall, L. E., Haggerty, D. J., Cooper, J. T., Golden, C. J., & Dornheim, L. (1998). *Development and validation of a measure of emotional*

Schutz, W. (1958). *FIRO: A Three Dimensional Theory of Interpersonal Behaviour*. Rinehart.

Schutz, W. C. (1958). *FIRO: A three-dimensional theory of interpersonal behavior*. Rinehart.

Selvaraj, K., & Lakshmanan, S. (2021). The Machine learning for predictive maintenance in supply chain management. Journal of Artificial intelligence and Machine Learning, 1(1), 9-15.

Senhaji-Mouhaddib, I., Bouzakri, S., & Lagrat, I. (2024, May). Advanced Autonomous Navigation Technologies for UAVs: Challenges and Opportunities," *2024 4th International Conference on Innovative Research in Applied Science, Engineering and Technology (IRASET)*, FEZ, Morocco, 1-7, https://doi.org/DOI: 10.1109/IRASET60544.2024.10548645

Shah, A. (2021). *AI and the Future of Manufacturing*. Elsevier.

Shan, T., Tang, W., Dang, X., Li, M., Yang, F., Xu, S., & Wu, J. (2020). Study on a fast solver for Poisson's equation based on deep learning technique. *IEEE Transactions on Antennas and Propagation, 68*(9), 6725–6733. DOI: 10.1109/TAP.2020.2985172

Sharma, R., Raissi, M., & Guo, Y. B. (2024). Physics-Informed Machine Learning for Smart Additive Manufacturing. arXiv preprint arXiv:2407.10761.

Shi, H. Y., Wang, S., & Wang, P. Y. (2024). From simulation to reality: CFD-ML-driven structural optimization and experimental analysis of thermal plasma reactors. *Journal of Environmental Chemical Engineering, 12*(3), 112998. DOI: 10.1016/j.jece.2024.112998

Shirvani, A., Nili-Ahmadabadi, M., & Ha, M. Y. (2023). Machine learning-accelerated aerodynamic inverse design. *Engineering Applications of Computational Fluid Mechanics, 17*(1), 2237611. DOI: 10.1080/19942060.2023.2237611

Silva, R. G., Sousa, A. C. M., & Lima, M. R. (2020). Automated drone inspections in aerospace: Challenges and future directions. *Aerospace Science and Technology, 106*, 106078. DOI: 10.1016/j.ast.2020.106078

Singh, A. K., & Patel, D. (2020, January). Optimization of Air Flow Over a Car by Wind Tunnel. In *2020 International Conference on Computation, Automation and Knowledge Management (ICCAKM)* (pp. 1-4). IEEE. DOI: 10.1109/ICCAKM46823.2020.9051457

Singh, M., & Khan, S. A. L. A. (2024). Advances in Autonomous Robotics: Integrating AI and Machine Learning for Enhanced Automation and Control in Industrial Applications. *International Journal for Multidimensional Research Perspectives*, *2*(4), 74–90. DOI: 10.61877/ijmrp.v2i4.135

Singh, V., & Patel, S. (2021). Implementation of AI-driven quality control in aerospace assembly lines. *Journal of Manufacturing Science and Engineering*, *143*(6), 061008.

Sivabalakrishnan, M., Patel, R., & Lee, J. (2020). IoT-based additive manufacturing system for customized production. *Journal of Manufacturing Systems*, *56*, 345–356.

Skydio. (2022). *Skydio autonomous flight systems*. https://www.skydio.com

Smith, A., & Jones, B. (2021). High-precision 3D reconstruction for small-to-medium-sized objects utilizing line-structured light scanning: A review. *Optical Engineering (Redondo Beach, Calif.)*, *60*(11), 110901.

Snell, J., Swersky, K., & Zemel, R. "Prototypical networks for few-shot learning" Advances in Neural Information Processing Systems (NeurIPS), 2017.

Sodiya, E. O., Jacks, B. S., Ugwuanyi, E. D., Adeyinka, M. A., Umoga, U. J., Daraojimba, A. I., & Lottu, O. A. (2024). Reviewing the role of AI and machine learning in supply chain analytics. *GSC Advanced Research and Reviews*, *18*(2), 312–320. DOI: 10.30574/gscarr.2024.18.2.0069

South African National Space Agency (SANSA). (2021). Annual report 2020/2021. https://www.sansa.org.za

South African National Space Agency (SANSA). (2021). *South African satellite missions*. https://www.sansa.org.za

Spalart, P., & Allmaras, S. (1992, January). A one-equation turbulence model for aerodynamic flows. In 30th aerospace sciences meeting and exhibit (p. 439). DOI: 10.2514/6.1992-439

Spalart, P. R. (2000). Strategies for turbulence modelling and simulations. *International Journal of Heat and Fluid Flow, 21*(3), 252–263. DOI: 10.1016/S0142-727X(00)00007-2

Spitzberg, B. H. (2015). Interpersonal Communication Competence and social skills. *The International Encyclopedia of Communication*, 1–8. DOI: 10.1002/9781405186407.wbieci078.pub2

Spitzberg, B. H. (1983). Communication competence as knowledge, skill, and Impression. *Communication Education*, *32*(3), 323–329. DOI: 10.1080/03634528309378550

Spitzberg, B. H., & Cupach, W. R. (1984). *Interpersonal communication competence* (Vol. 4). SAGE Publications, Incorporated.

Stone, J. (2016). Interpersonal Communication and Identity. In *Communication Competence: Creating Satisfying Interpersonal Relationships* (pp. 93–121). essay, University of Minnesota Libraries Publishing.

Stutts, L. (2022). Increasing Self-Compassion: Review of the Literature and Recommendations. *Journal of undergraduate neuroscience education: JUNE: a publication of FUN. Faculty for Undergraduate Neuroscience, 20*(2), A115–A119. DOI: 10.59390/WSZK3327 PMID: 38323068

Sukhbaatar, S., Szlam, A., Welson, J., & Fergus, R. "End-To-End Memory Networks", arXiv:1503.08895, doi:/arXiv.1503.08895, 2015DOI: 10.48550

Sullivan, H. S. (1953). *The Interpersonal Theory of Psychiatry*. W. W. Norton.

Sun, D., Zheng, S., Li, Y., & Jia, M. (2021). A machine learning approach to air traffic flow management. *Transportation Research Part C, Emerging Technologies*, *129*, 103254. DOI: 10.1016/j.trc.2021.103254

Sun, H., Li, S., Zheng, X., & Lu, X. (2020). Remote sensing scene classification by gated bidirectional network. *IEEE Transactions on Geoscience and Remote Sensing*, *58*(1), 82–96. DOI: 10.1109/TGRS.2019.2931801

Sunil, M. (2013). 8. Evolution of the sectoral system of innovation of India's aeronautical industry. *International Journal of Technology and Globalisation*, *7*(1/2), 92. Advance online publication. DOI: 10.1504/IJTG.2013.052033

Sun, K., Wang, W., Cheng, R., Liang, Y., Xie, H., Wang, J., & Zhang, M. (2024). Evolutionary generative design of supercritical airfoils: An automated approach driven by small data. *Complex & Intelligent Systems*, *10*(1), 1167–1183. DOI: 10.1007/s40747-023-01214-0

Tao, R., Tian, F., Huang, H., & Yang, H. (2023). AI-accelerated computational framework for fluid–structure interaction analysis in aerospace applications.

Tavares, S. M. O., Ribeiro, J. A., Ribeiro, B. A., & de Castro, P. M. S. T.Sérgio. (2024, March 22). Mo, Tavares., João, A., Ribeiro., Bruno, A., Ribeiro., Paulo, M.S.T., de, Castro. (2024). 4. Aircraft Structural Design and Life-Cycle Assessment through Digital Twins. *Designs*, *8*(2), 29. Advance online publication. DOI: 10.3390/designs8020029

Thangamani, R., Suguna, R. K., & Kamalam, G. K. (2024). Drones and Autonomous Robotics Incorporating Computational Intelligence. In: Prakash, K. B., Peddapelli, S. K., Tam, I. C. K., Woo, W. L., Jain, V. (eds) *Computational Intelligent Techniques in Mechatronics*, 243-296, John Wiley & Sons, Inc., Hoboken, NJ, US and Scrivener Publishing, Beverly, MA, US DOI: 10.1002/9781394175437.ch8

The Role of Digital Twins in AI-Driven QC. (2024).. . *Aerospace Engineering Review*, *55*(6), 310–328.

Tong, M. T. (2023, June). Aero-Engines AI-A Machine-Learning App for Aircraft Engine Concepts Assessment. In Turbo Expo: Power for Land, Sea, and Air (Vol. 86939, p. V001T01A017). American Society of Mechanical Engineers. DOI: 10.1115/GT2023-102024

Tong, P., Yang, X., Yang, Y., Liu, W., & Wu, P. (2023). Multi-UAV collaborative absolute vision positioning and navigation: A survey and discussion. *Drones (Basel)*, *7*(4), 261. DOI: 10.3390/drones7040261

Tseng, M.-H. (2023). GA-based weighted ensemble learning for multi-label aerial image classification using convolutional neural networks and vision transformers. *Machine Learning: Science and Technology*, *4*(4), 045045. DOI: 10.1088/2632-2153/ad10cf

Tyystjärvi, T., Fridolf, P., Rosell, A., & Virkkunen, I. (2024). Deploying Machine Learning for Radiography of Aerospace Welds. *Journal of Non-destructive Evaluation*, *43*(1), 24. DOI: 10.1007/s10921-023-01041-w

UNESCO. (2021). Science report: The race against time for smarter development. UNESCO Publishing. https://unesdoc.unesco.org/ark:/48223/pf0000377257

United Nations Development Programme (UNDP). (2020). *Human development report 2020: The next frontier—Human development and the Anthropocene*. UNDP. https://hdr.undp.org

Upadhyay, J., Rawat, A., & Deb, D. (2021). Multiple drone navigation and formation using selective target tracking-based computer vision. *Electronics (Basel)*, *10*(17), 2125. DOI: 10.3390/electronics10172125

Vignesh, P., Maheswari, R., Vijaya, P., & Vignesh, U. (2024). Machine Learning for Aerospace Object Categorization. In AI and Blockchain Optimization Techniques in Aerospace Engineering (pp. 164-180). IGI Global. DOI: 10.4018/979-8-3693-1491-3.ch008

Vinuesa, R., Hosseini, S. M., Hanifi, A., Henningson, D. S., & Schlatter, P. (2017). Pressure-gradient turbulent boundary layers developing around a wing section. *Flow, Turbulence and Combustion*, *99*(3-4), 613–641. DOI: 10.1007/s10494-017-9840-z PMID: 30069159

Vinyals, O., Blundell, C., Lillicrap, T., Kavukcuoglu, K., & Wierstra, D. "Matching networks for one shot learning", *Proceedings of the 30th International Conference on Neural Information Processing Systems*, pp. 3637–3645, 2016.

Vocate, D. R. (1994). *Intrapersonal communication: different voices, different minds*. Erlbaum.

Volpe, M. R., & Chandler, D. (1999). Resolving Conflicts in Institutions of Higher Education: Challenges for Pracademics. *CNCR-Hewlett Foundation Seed Grant White Papers*. 8. https://readingroom.law.gsu.edu/seedgrant/8

Vygotsky, L. S. (1987). Thinking and Speech. In R. W. Rieber, & A. S. Carton (Eds.), The Collected Works of L. S. Vygotsky (Vol. 1), Problems of General Psychology (pp. 39-285). New York: Plenum Press. (Original Work Published 1934)

Vygotsky, L. S. (1930/1978). *Mind in Society: The development of higher psychological processes*. Harvard University Press.

Wang, B., & Wang, J. (2021). Application of artificial intelligence in computational fluid dynamics. *Industrial & Engineering Chemistry Research*, *60*(7), 2772–2790. DOI: 10.1021/acs.iecr.0c05045

Wang, W., & Ma, J. (2024). A review: Applications of machine learning and deep learning in aerospace engineering and aero-engine engineering. *Advances in Engineering Innovation*, *6*(1), 54–72. DOI: 10.54254/2977-3903/6/2024060

Wang, Y., & Li, J. (2022). Real-time monitoring of aerospace manufacturing processes using machine learning. *International Journal of Advanced Manufacturing Technology*, *120*(9-12), 3456–3467.

Warey, A., Raul, V., Kaushik, S., Han, T., & Chakravarty, R. (2023). Generative Inverse Design of Aerodynamic Shapes Using Conditional Invertible Neural Networks. *Journal of Computing and Information Science in Engineering*, *23*(3), 031006. Advance online publication. DOI: 10.1115/1.4054715

Weller, H. G., Tabor, G., Jasak, H., & Fureby, C. (1998). A tensorial approach to computational continuum mechanics using object-oriented techniques. *Computers in Physics*, *12*(6), 620–631. DOI: 10.1063/1.168744

Wong, K. (2023). *Data-Driven Quality Control in Aerospace Engineering*. Springer.

World Bank. (2021). World development indicators. https://data.worldbank.org

Wu, J., Xiao, H., Sun, R., & Wang, Q. (2019). Reynolds-averaged Navier–Stokes equations with explicit data-driven Reynolds stress closure can be ill-conditioned. *Journal of Fluid Mechanics*, *869*, 553–586. DOI: 10.1017/jfm.2019.205

Xin, L., Tang, Z., Gai, W., & Liu, H. (2022). Vision-based autonomous landing for the UAV: A review. *Aerospace (Basel, Switzerland)*, *9*(11), 634. DOI: 10.3390/aerospace9110634

Xue, T., Chen, L., Hu, J., & Zhou, Q. (2023). A variable-fidelity multi-objective optimization method for aerospace structural design optimization. *Engineering Optimization*, *55*(7), 1133–1148. DOI: 10.1080/0305215X.2022.2061964

Yang, Y., Wei, K., Kang, R., & Wang, S. (2018). Multi-objective chance constrained programming of spare parts based on uncertainty theory. *IEEE Access : Practical Innovations, Open Solutions*, *6*, 50049–50054. DOI: 10.1109/ACCESS.2018.2860252

Yarlanki, S., Rajendran, B., & Hamann, H. (2012, May). Estimation of turbulence closure coefficients for data centers using machine learning algorithms. In 13th InterSociety Conference on Thermal and Thermo-mechanical Phenomena in Electronic Systems (pp. 38-42). IEEE. DOI: 10.1109/ITHERM.2012.6231411

Zamaraev, K. I. (1996). Photocatalysis: State of the art and perspectives. *Studies in Surface Science and Catalysis*, *101*, 35–50.

Zhang, Y., Liu, J., Shi, S., & Wang, H. (2021). Machine learning-assisted prediction of ice accretion on aircraft surfaces.

Zhang, T., Li, H., & Zhao, Y. (2021). Accurate and reliable sealant inspection for aircraft fuel tanks based on 3D point cloud. *Journal of Aerospace Engineering*, *34*(6), 04021085.

Zhang, Y., He, J., & Wei, B. (2020). Machine learning in nondestructive testing: Advances and perspectives. *NDT & E International, 116*, 102288.

Zhang, Y., Yang, G., & Zhang, Y. (2019). Intelligent fault diagnosis of aerospace systems using machine learning: Review and perspectives. *Chinese Journal of Aeronautics, 32*(1), 213–236. DOI: 10.1016/j.cja.2018.09.012

Zhan, Z., & Li, H. (2021). A novel approach based on the elastoplastic fatigue damage and machine learning models for life prediction of aerospace alloy parts fabricated by additive manufacturing. *International Journal of Fatigue, 145*, 106089. Advance online publication. DOI: 10.1016/j.ijfatigue.2020.106089

Zhao, R., Yan, R., Chen, Z., Mao, K., Wang, P., & Gao, R. X. (2019). Deep learning and its applications to machine health monitoring: A survey. *Mechanical Systems and Signal Processing, 115*, 213–237. DOI: 10.1016/j.ymssp.2018.05.050

Zhizhong, C., Hongqing, L., Zengcong, L., Yan, C., Jie, C., & Xiaoqi, L. (2023). Structural lightweight design and experimental validation for aerospace sealed cabin. *Frontiers of Mechanical Engineering, 9*, 1265734. DOI: 10.3389/fmech.2023.1265734

Zhong, J., Li, M., Chen, Y., Wei, Z., Yang, F., & Shen, H. (2024). A safer vision-based autonomous planning system for quadrotor UAVs with dynamic obstacle trajectory prediction and its application with LLMs. *IEEE/CVF Winter Conference on Applications of Computer Vision Workshops (WACVW)*, Waikoloa, HI, USA, 920-929, https://doi.org/DOI: 10.1109/WACVW60836.2024.00131

Zhou, B., Lapedriza, A., Khosla, A., Oliva, A., & Torralba, A. (2018). Places: A 10 million Image Database for Scene Recognition. *IEEE Transactions on Pattern Analysis and Machine Intelligence, 40*(6), 1452–1464. DOI: 10.1109/TPAMI.2017.2723009 PMID: 28692961

Zhou, H. R., Yang, H., Li, H. Q., Ma, Y. C., Yu, S., Shi, J., Cheng, J., Gao, P., Yu, B., Miao, Z., & Wei, Y. P. (2024). Advancements in machine learning for material design and process optimization in the field of additive manufacturing. *China Foundry*, *21*(2), 101–115. DOI: 10.1007/s41230-024-3145-3

Zhu, L., Zhang, W., Kou, J., & Liu, Y. (2019). Machine learning methods for turbulence modelling in subsonic flows around airfoils. *Physics of Fluids*, *31*(1), 015105. DOI: 10.1063/1.5061693

Zody, R. E. (1977). Academic/practitioner interchanges: Suggested guidelines. *Public Administration Review*, *37*(5), 572. DOI: 10.2307/974719

Zourbanos, N., Hatzigeorgiadis, A., & Theodorakis, Y. (2007). A preliminary investigation of the relationship between athletes' self-talk and coaches' behaviour and statements. *International Journal of Sports Science &amp. International Journal of Sports Science & Coaching*, *2*(1), 57–66. DOI: 10.1260/174795407780367195

About the Contributors

Venkata Tulasiramu Ponnada stands as a luminary in the dynamic field of computer science, bringing with them a wealth of expertise, innovation, and scholarly accomplishments. As a distinguished professor, researcher, and mentor, Dr. Ponnada has continually pushed the boundaries of knowledge, contributing significantly to both academia and industry. With a career spanning 17 years, Dr. Ponnada has established themselves as a thought leader in safety critical system(aerospace and Healthcare) [implementation and machine learning, garnering recognition for their pioneering research and groundbreaking contributions. Their work has not only advanced the theoretical foundations of computer science but has also translated into real-world applications, shaping the technological landscape in profound ways. As an educator, Dr. Ponnada is renowned for their passion for teaching and commitment to fostering the next generation of computer scientists. Their engaging lectures, innovative teaching methods, and mentorship have inspired countless students to explore the frontiers of technology and pursue excellence in their academic pursuits. In addition to their academic endeavors, Dr. Ponnada is actively engaged in collaborative research projects, interdisciplinary initiatives, and industry partnerships. Their collaborative spirit and interdisciplinary approach have led to transformative breakthroughs, bridging the gap between theory and practice and driving innovation across diverse domains. Beyond their professional achievements, Dr. Ponnada is known for their dedication to service and leadership within the academic community. Whether serving on editorial boards, organizing conferences, or advocating for diversity and inclusion in STEM fields, they are committed to advancing the field of computer science and fostering a supportive and inclusive environment for all. In summary, Dr. Ponnada embodies the spirit of excellence, innovation, and leadership in computer science. Their contributions have had a lasting impact on the field, inspiring colleagues, students, and industry professionals alike. As we look to the future of technology, Dr. Ponnada continues to lead the way, shaping the course of computer science and paving the path for generations to come.

***

M.L Azad is Teaching experience 25 Year and research area expert electrical engineering specialization of electric grid and solar power.

Dipanjan Das is an MCA graduate from Brainware University, Class of 2023. I have a strong interest in artificial intelligence and machine learning-driven technologies. My academic journey has equipped me with a solid foundation in computer science, and I am passionate about exploring innovative solutions using AI/ML. I am eager to contribute to research and development in this field, aiming to solve real-world problems through intelligent systems. I continuously strive to enhance my skills and knowledge, with the goal of building a meaningful career in the ever-evolving landscape of AI and machine learning.

Boopathy Govindarajan earned his undergraduate degree in mechanical engineering from University of Madras, his master's degree in aeronautical engineering from Madras Institute of Technology, and his doctorate from the Department of Aeronautical Engineering, St. Peter's Institute of Higher Education and Research, Chennai. At Vel Tech University, a professorship is currently held in the department of aeronautical engineering. In addition to guiding 5 Ph.D Scholars, guided 8 PG & 90 UG scholars on their significant projects, he has written in 30 international publications. He is a Life Member of the Society of Automotive Engineers INDIA and the Aeronautical Society of India.

Prasad G is currently working as faculty in the department of Aerospace Engineering Chandigarh University, India. He has 9 years of experience in Teaching and research. He received his Bachelor of Engineering and Master of Engineering degrees with first class and distinction and Doctor of Philosophy in the field of Aeronautical Engineering from the Anna University, Chennai, India. Prof. Prasad G is the author of over 30 technical publications, 1 Patent Published and 1 Book published. His research interests include Unmanned Aerial Vehicles, Aircraft Structures, Computational Fluid Dynamics and Interdisciplinary research. He has completed three funded project sponsored by The Institution of Engineers (India) and Tamilnadu State Council for Science and Technology. Awarded Indian National Science Academy (INSA) Visiting Scientist Programme 2019 and Awarded Science Academies' Summer Research Fellowship Programme (SRPF) 2019. Further he is SPOC of NPTEL- IIT Madras, SPOC of ISRO IIRS outreach coordinator and SPOC for NASA International SpaceApps Challenge. Further key positions such as coordinator of Board of Studies, Outcome Based Education, Project Coordinator, Research Coordinator

Balaji Ganesan is a prominent aerodynamicist and professor at the Hindustan Institute of Technology and Science (HITS) in Chennai, India. With a specialization in aerodynamics, fluid dynamics, and propulsion systems, he has established himself as a leader in aerospace engineering research and education. Dr. Balaji earned his doctoral degree in Aerospace Engineering, focusing on high-speed aerodynamics, and has extensive experience in both academic and applied research within the field. At HITS, Dr. Balaji is involved in teaching, mentoring, and guiding students in aerospace and mechanical engineering programs. His research encompasses a wide range of topics within aerodynamics, including computational fluid dynamics (CFD), experimental fluid mechanics, and aircraft design. Dr. Balaji has published numerous research articles in reputed international journals and has presented at various prestigious conferences worldwide. His work often addresses critical issues in aerodynamics and propulsion, aiming to improve performance, efficiency, and safety in aerospace applications. Dr. Balaji also collaborates with national and international aerospace organizations, contributing to projects that advance aeronautical science and engineering. He is passionate about developing innovative, sustainable solutions for aerospace challenges and is dedicated to inspiring the next generation of engineers through hands-on research and immersive learning experiences.

Aadya Gupta is an aerospace engineering student with a strong foundation in aerodynamics, propulsion, and structural analysis. My academic and project experiences have honed my problem-solving skills and passion for innovation in aviation and space technology. I'm particularly interested in spacecraft design and drone . I am eager to apply my knowledge through internships and research opportunities, contributing to advancements in the aerospace industry.

Kaushik Kumar holds a Ph.D. in Engineering from Jadavpur University, India, an MBA in Marketing Management from Indira Gandhi National Open University, India and a Bachelor of Technology from Regional Engineering College (Now National Institute of Technology), Warangal, India. For 11 years, he worked in a manufacturing unit of Global repute. He is currently working as a Professor in the Department of Mechanical Engineering, Birla Institute of Technology, Mesra, Ranchi, India. He has 23 Years of Teaching and Research Experience. His research interests include Composites, Optimization, Non-Conventional Machining, CAD / CAM, Rapid Prototyping and Quality Management Systems towards product development for societal and industrial usage and has received 40+ patents for them most of which are donated to Start-ups and Industrial fraternity through Jharkhand Small Industries Association (JSIA). He has published 60+ Books (They are referred as Text Books and Reference Books by 40+ Universities / Institutes in their academic curriculum), 100+ Book Chapters and 200+ Research Papers in peer-reviewed reputed national and international journals. Kaushik has also served as Editor–in–Chief, Series Editor, Guest Editor, Editor, Editorial Board Member and Reviewers for International and National Journals. He has been felicitated with many awards and honours including Distinguished Alumnus Award for Professional Excellence 2023 under Academic and Research from National Institute of Technology, Warangal, India. He has also received Sponsored Research and Consultancy Projects of more than 1 Crore from Govt. of India and abroad. Kaushik has delivered expert lectures as Keynote Speaker at International & National Conferences, Resource Person at Various Workshops, FDP's and Short-

Term Courses. He has guided many students of Doctoral, Masters and Under Graduate programmes of his home and other institutions in India and abroad. He also has served as Reviewer and Examiner of Doctoral and Masters dissertation for institutes in India and abroad.

Snehal Malik is an aspiring Aerospace Engineer and a student of the universe, my passion for space exploration knows no bounds. My love for the universe drives me to pursue a career where I can contribute to the advancement of aerospace technology and exploration. My dedication to understanding the complexities of the universe and my commitment to excellence makes me standout with a strong desire to make a meaningful impact in the field of aerospace engineering.

Yash Nigam is a passionate aerospace engineering student from Sitapur, Uttar Pradesh. Currently pursuing my degree at Chandigarh University, I'm dedicated to exploring the exciting world of aviation and space exploration. With a strong drive and ambition, I'm eager to contribute to the advancement of aerospace technology and achieve my career goals in this field.

Sivaprakasam Palani received the B.E. M.E., and Ph.D. degree in Faculty of Mechanical Engineering from the College of Engineering, Guindy, Anna University, Chennai, India. He is currently working as an Associate Professor with Addis Ababa Science and Technology University, Ethiopia. He has been supervising Master and Ph.D. students in the area of Machining, Composite materials and Surface Engineering. He has 15 years of teaching and research experience. He has published and presented over 100 technical papers in international journals, conferences, and book chapters. His research interests include Micro/nano manufacturing, Machining, Composite materials, Surface Engineering, and Optimization techniques.

Balaji PS received his PhD in vibration control from Curtin University, Australia and Master in Technology in Mechanical Engineering from National Institute of Technology Allahabad. He did his Bachelors of Engineering in Aeronautical Engineering from Anna University Chennai. His area of research is Vibration, Composite materials and Advanced machining Process. He is actively involved in the consultancy projects on vibrations and stress analysis of mechanical components. He has also undertaken funded projects in the field of vibrations and composite materials. He is currently, associated with National Institute of Technology Rourkela in the Department of Mechanical Engineering. He has also authored patents on the vibration control devices. He also a reviewer for reputed International and National journals.

Vinob Chander R is an Associate Professor with over 21 years of experience spanning academia and industry. He holds a Ph.D. in Internet of Things, M.E in CSE, and B.E. in Computer Engineering. He is a Life Member of ISTE, Member of IETE, and Member of IAENG. His research interests include AI applications in aerospace, machine learning for predictive maintenance, and IoT-based systems. Dr. Vinob has authored multiple publications in reputed international journals and conferences, focusing on AI, IoT, and smart environments. He has led several funded research projects, including work on smart home systems and AI-driven applications. His expertise in software development, particularly in areas such as data science, web technologies, and embedded systems, provides a strong foundation for exploring AI applications in aerospace. As a Sr. Manager and corporate trainer, Dr. Vinob has designed courses on embedded systems, IoT, and data science for major tech companies. He has also supervised numerous projects related to AI and machine learning applications, demonstrating his practical approach to emerging technologies. Dr. Vinob's combination of academic research, industry experience, and hands-on project work positions him uniquely to address the cutting-edge applications of AI in aerospace design optimization and predictive maintenance.

Chikesh Ranjan is currently serving as a Project Engineer for Project SwaYaan – Capacity Building in Drone/UAS at the Department of Mechanical Engineering, NIT Rourkela, India. He holds a PhD in Mechanical Engineering from the Department of Mechanical Engineering at the Birla Institute of Technology, Mesra, Ranchi, Jharkhand, India, where he also completed his Master of Engineering (M.E.) in Mechanical Engineering, specializing in the Design of Mechanical Equipment. Additionally, he earned a Bachelor of Engineering (B.E.) in Mechanical

Engineering from the Marathwara Institute of Technology, Aurangabad, Maharashtra, India, graduating with First Class with Distinction. Chikesh Ranjan has made significant contributions to the field of mechanical engineering through his research, having published 17 journal papers, authored 12 books, and written 16 book chapters. He has been actively involved in organizing 10 workshops and coordinating 13 events, in addition to participating in 31 training programs and attending 34 national and international workshops, seminars, and webinars. His innovative work has been recognized with 13 design patents and 8 foreign patents, underscoring his commitment to advancing mechanical design and technology. With 9 years of teaching and research experience, Chikesh has built a global reputation in his areas of interest, which include composites, non-conventional machining, CAD/CAM, 3D printing, robotics and drone technology. He is a proud member of the Institution of Engineers, India, and has received 7 prestigious awards and honors for his outstanding contributions to the field of Mechanical Engineering.

J. Srinivas is currently working as a Professor in Mechanical Engineering, NIT Rourkela. He guided several Ph.D. and M.Tech students in the field of dynamics and control. He published several research articles in reputed journals and conferences. He authored/co-authored around twelve books in different domains including neural networks, computer aided design and manufacturing and experimental techniques in Mechanical Engineering. His topics of interest include: robotics, drone technology and control techniques.

Kumaran T is working as an assistant professor in department of Aeronautical engineering.

Index

A

Additive Manufacturing 6, 19, 38, 39, 40, 41, 42, 59, 67, 68, 85, 164, 189, 190, 193, 194
Aerospace 1, 2, 3, 4, 5, 6, 7, 8, 10, 11, 12, 13, 14, 15, 16, 17, 18, 19, 20, 21, 22, 23, 24, 25, 26, 27, 28, 29, 30, 31, 32, 33, 34, 35, 36, 37, 38, 39, 40, 42, 43, 44, 45, 47, 48, 49, 52, 55, 57, 58, 59, 60, 61, 62, 63, 65, 66, 67, 68, 69, 70, 71, 73, 74, 75, 76, 77, 78, 79, 80, 81, 82, 83, 84, 85, 107, 109, 147, 150, 153, 158, 159, 161, 162, 163, 164, 165, 166, 167, 168, 169, 172, 173, 174, 175, 176, 177, 178, 179, 180, 181, 182, 183, 184, 185, 186, 187, 188, 189, 190, 191, 192, 193, 194, 195, 225, 226, 227, 228, 229, 230, 231, 233, 234, 235, 236, 237, 238, 239, 240, 241, 242
aerospace innovation 61, 225, 226, 229, 230, 231, 236
aerospace manufacturing 4, 6, 14, 15, 16, 17, 18, 19, 20, 21, 33, 37, 45, 58, 67, 161, 162, 163, 164, 165, 167, 168, 169, 172, 173, 175, 176, 181, 182, 183, 184, 185, 186, 189, 191, 193, 194, 195
Aircraft 3, 4, 5, 6, 7, 8, 9, 10, 13, 14, 16, 17, 18, 22, 26, 27, 28, 29, 31, 32, 33, 34, 37, 39, 42, 43, 44, 45, 46, 47, 48, 49, 50, 58, 59, 63, 64, 65, 66, 69, 70, 71, 73, 74, 75, 77, 78, 80, 84, 136, 148, 149, 157, 159, 162, 167, 168, 172, 174, 175, 176, 177, 178, 183, 187, 191, 192, 195, 225, 226, 227, 228, 229, 231, 232, 233, 234, 235, 236, 238, 241
anomaly detection 4, 58, 72, 73, 74, 75, 81, 162, 163
Artificial Intelligence 2, 14, 30, 32, 33, 34, 35, 41, 44, 46, 47, 48, 49, 55, 61, 77, 78, 107, 137, 159, 162, 163, 172, 188, 226, 230
automation 5, 7, 11, 12, 13, 20, 22, 38, 40, 41, 45, 48, 63, 67, 70, 74, 76, 77, 81, 162, 163, 183, 188, 189, 194, 232
autonomous flight 27, 43, 100, 153, 242

C

coaching 197, 198, 199, 200, 202, 203, 204, 205, 206, 207, 208, 209, 210, 211, 212, 213, 214, 215, 216, 217, 218, 219, 220, 221, 222, 224
communication 30, 80, 108, 197, 198, 199, 202, 203, 204, 205, 206, 207, 208, 209, 210, 211, 213, 214, 215, 216, 217, 219, 221, 222, 223, 226, 230, 237
Computational Fluid Dynamics 2, 3, 45, 46, 54, 55, 64, 135, 136, 137, 138, 142, 151, 154, 155, 157, 158, 159
Computer vision 59, 68, 87, 88, 89, 90, 91, 97, 98, 99, 100, 101, 102, 103, 104, 105, 108, 109, 132, 147, 161, 162, 163, 164, 165, 170, 172, 175, 176, 177, 178, 179, 183, 185, 187, 191, 194, 195, 232, 233
Computer Vision for Drone Navigation 98
Convolutional Neural Networks 16, 17, 18, 88, 101, 102, 104, 113, 117, 133, 163, 165, 174, 190, 191, 233

D

deep learning 17, 24, 26, 37, 42, 43, 46, 54, 55, 75, 88, 92, 93, 94, 95, 100, 103, 105, 108, 113, 117, 124, 136, 138, 158, 162, 165, 174, 178, 182, 189, 190, 191, 192, 193, 194, 228, 232, 233, 242
defect detection 7, 17, 43, 72, 162, 163, 165, 170, 172, 173, 175, 176, 180, 183, 184, 185, 187, 188, 190, 191, 192, 194
Design 1, 2, 3, 5, 6, 7, 8, 9, 10, 11, 12, 13, 14, 15, 16, 21, 22, 23, 25, 28, 31, 32, 34, 36, 37, 38, 39, 40, 41, 42, 43, 47, 57, 58, 59, 60, 61, 62, 63, 64, 65, 66, 77, 78, 79, 80, 81, 82, 83, 84, 85, 117, 119, 121, 136, 137, 139, 140, 142,

146, 148, 149, 150, 152, 154, 177, 182, 193, 197, 225, 226, 227, 231, 236, 238, 240
Design Optimization 3, 5, 9, 12, 13, 40, 64, 85, 146
Drones and Feature Detection 98

E

edge AI 104, 180, 184, 192, 237
emerging economies 227, 231, 234, 235
emotional intelligence 199, 200, 208, 209, 210, 214, 215, 216, 217, 218, 219, 220, 221

G

generative design 11, 12, 22, 31, 36, 40, 47, 65, 66, 79, 81, 84, 238, 240

I

interpersonal 197, 198, 199, 200, 202, 203, 204, 205, 206, 207, 208, 209, 210, 211, 213, 214, 215, 216, 217, 218, 219, 222, 223
intrapersonal 197, 198, 199, 200, 205, 206, 207, 208, 209, 210, 211, 213, 214, 216, 217, 218, 219, 221, 222, 223

M

Machine Learning 1, 2, 3, 4, 5, 6, 7, 12, 13, 14, 17, 20, 22, 24, 25, 27, 28, 29, 36, 37, 38, 39, 40, 41, 42, 43, 44, 45, 47, 48, 49, 52, 53, 56, 57, 58, 59, 60, 61, 63, 64, 65, 66, 67, 68, 69, 70, 72, 73, 74, 75, 76, 77, 78, 79, 80, 81, 82, 83, 84, 85, 88, 100, 102, 115, 133, 135, 136, 137, 138, 139, 141, 142, 143, 144, 146, 147, 148, 149, 150, 151, 152, 153, 154, 155, 156, 157, 159, 161, 162, 163, 164, 165, 170, 172, 173, 174, 175, 176, 177, 179, 181, 182, 183, 185, 188, 190, 191, 195, 225, 226, 227, 228, 229, 230, 231, 232, 233, 234, 235, 236, 238, 241, 242
Machine Learning in Aerospace 1, 7, 14, 63, 67, 165, 175, 177, 185, 236
Modeling 38, 54, 62, 65, 66, 84, 117, 136, 137, 138, 158, 188, 190

O

Oriented FAST and Rotated BRIEF 98, 99

P

Predictive Maintenance 3, 6, 16, 17, 19, 26, 30, 31, 36, 37, 38, 41, 43, 44, 46, 58, 62, 69, 71, 72, 74, 75, 77, 81, 85, 136, 142, 157, 162, 165, 171, 173, 179, 180, 185, 187, 191, 227, 235
process optimization 15, 20, 41, 42, 67, 162, 163, 179, 185, 188, 190

Q

Quality Assurance 2, 48, 71, 74, 163, 185, 189
quality control 17, 18, 32, 45, 47, 59, 67, 68, 71, 76, 81, 155, 161, 162, 163, 164, 165, 166, 167, 168, 169, 172, 173, 174, 179, 180, 181, 182, 183, 184, 185, 188, 189, 190, 194, 195
quantum computing 6, 52, 58, 60, 63, 77, 78, 79, 80, 81, 240, 242

S

Scale-Invariant Feature Transform 98, 99
self-talk 199, 200, 205, 206, 207, 208, 215, 216, 221, 224
Simulation 5, 11, 26, 45, 46, 48, 49, 52, 53, 55, 59, 78, 136, 137, 138, 142, 143, 144, 145, 148, 150, 151, 152, 154, 157, 158, 236

T

talent development 235
Technology 1, 2, 4, 19, 22, 24, 25, 27, 28, 29, 30, 31, 34, 36, 37, 38, 41, 44, 49, 50, 57, 60, 61, 63, 67, 69, 80, 83, 84,

85, 87, 96, 99, 104, 105, 108, 133, 151, 164, 184, 185, 188, 193, 195, 225, 226, 229, 231, 232, 233, 234, 235, 238, 239, 240, 241, 242